T0205476

Snow Optics

Alexander Kokhanovsky

Snow Optics

 Springer

Alexander Kokhanovsky
A Leonardo and Thales Company
Telespazio Belgium
Darmstadt, Germany

ISBN 978-3-030-86591-7   ISBN 978-3-030-86589-4   (eBook)
https://doi.org/10.1007/978-3-030-86589-4

This Springer imprint is published by the registered company Springer Nature Switzerland AG
The registered company address is: Gewerbestrasse 11, 6330 Cham, Switzerland

Under the microscope, I found that snowflakes were miracles of beauty; and it seemed a shame that this beauty should not be seen and appreciated by others. Every crystal was a masterpiece of design, and no one design was ever repeated. When a snowflake melted, that design was forever lost. Just that much beauty was gone, without leaving any record behind.

Wilson Bentley (1925)

Preface

Permanent and seasonal snow covers large areas of terrestrial surface. Unlike liquid water, snowpack is a highly reflective surface in the visible region of electromagnetic spectrum. The snowpack influences a variety of physical, chemical, biological, hydrological, and geological processes. Changes of snow extent and albedo influence climate on our planet. Therefore, it is of great importance to study snow properties using ground, airborne, and satellite instrumentation.

This book is an introduction to snow optics. The focus is on the studies of radiative transfer in snowpack and the solution of inverse radiative transfer problems related to the determination of snow properties from reflected solar radiation.

My research in snow optics has been inspired by Eleonora Zege, whom I am indebted for the introduction to this research filed at the end of last century. My snow optics and remote sensing research greatly benefited from cooperation with many colleagues around the world including John Burrows, Vladimir Rozanov, Leonid Dombrovsky, Wolfgang von Hoyningen-Huene, Teruyuki Nakajima, Teruo Aoki, Masahiro Hori, Georg Heygster, Takashi Nakajima, Jason Box, Ghislain Picard, Marie Dumont, Maxim Lamare, Florent Domine, Baptiste Vandecrux, Carsten Brockmann, Olaf Danne, Simon Gascoin, Marco Tedesco, Alexei Lyapustin, Sergey Korkin, Petri Raisanen, Jouni Peltoniemi, Andreas Macke, Stephen Warren, Richard Brandt, Tom Grenfell, Don Perovich, Odelle Hadley, Harendra Negi, Iosif Katsev, Alexander Prikhach, Biagio Di Mauro, and Michael Mishchenko.

While writing this book, my research was supported by the European Space Agency and Japan Aerospace Exploration Agency.

I thank my family for having endured my preoccupation with working on the book over the past several years.

Darmstadt, Germany Alexander Kokhanovsky
June 2021

Contents

Chapter 1
Microphysics and Geometry of Snowpack

1.1 Ice Grains in Snow: Size, Density and Shape

Snow is composed of close-packed ice crystals suspended in air. Most of crystals in snow have irregular shapes. Although falling snow crystals often have a striking symmetry (Bentley and Humphreys 1962). There are multiple ways to classify the dimension of crystals. Most common approach is to measure and report the largest dimension d of a crystal. Depending on the value of d, a given snowpack can be classified in six categories given in Table 1.1 The first three categories are most common. Snow crystals originate from atmosphere. Very fine size of crystals may serve as an indication of a recent snowfall because crystals grow with time. Snow metamorphism is driven by gradients in vapor pressure, which in turn are driven by temperature gradients. Small temperature gradients (less than 10 degree per meter) result in small vapor pressure gradients and slow grain growth within the snowpack. With time, crystals become more rounded. This is because the vapor diffusion within the snowpack causes a loss of mass from points on individual snow grains to gains in mass in hollows. The large temperature gradient induces a large gradient in vapor pressure, such that water vapor moves from warmer areas of the snowpack with relatively higher vapor pressures across pore spaces to colder areas of the snowpack with lower vapor pressure. These conditions produce angular or faceted grains, which may later develop steps and striations on their surface, resulting in cup-shaped crystals with a hollow centre that generally range in size from 3 to 8 mm. Under very favourable conditions, individual grains can be larger than 15 mm. The ice crystals in snow can be solid, hollow, broken, abraded, partly melted, rounded or angular. The surface of crystals can be rimed, stepped or striated. The rounded facets can be present as well. The crystals can be bounded, unbounded, clustered, or arranged in columns. The ice crystals of snow can be randomly distributed or partially (say, in horizontal plane) oriented. The diversity of crystals in snow is demonstrated in Fig. 1.1a, b.

Each crystal in snowpack differs from other crystals both in terms of shape and size. Therefore, the precise characterization of the snow microstructure is much

© Springer Nature Switzerland AG 2021
A. Kokhanovsky, *Snow Optics*,
https://doi.org/10.1007/978-3-030-86589-4_1

Table 1.1 The international classification of snow grain sizes (the largest dimension of ice grains) (Fierz et al. 2009)

Type	Term	Predominant maximal dimension of crystals (mm)
1	Very fine	<0.2
2	Fine	0.2–0.5
3	Medium	0.5–1.0
4	Coarse	1.0–2.0
5	Very coarse	2.0–5.0
6	Extreme	>5.0

more complex as compared, e.g., to the case of water clouds composed of spherical droplets. In case of spherical polydispersions one can introduce the average radius of particles

$$\bar{a} = \int_0^\infty a f(a)\, da, \tag{1.1}$$

where a is the radius of particles, $f(a)$ is the particle size distribution normalized as follows:

$$\int_0^\infty f(a)\, da = 1. \tag{1.2}$$

In earlier works related to snow optics, it has been a standard approach to model the snow optical properties assuming the spherical shape of particles. This made it possible to simplify modelling and derive useful theoretical results for snow radiative properties on the basis of well-known results of Maxwell theory as applied to spherical scatterers. The snow microstructure has been represented by a single parameter called the effective grain radius:

$$a_{\text{ef}} = \frac{\int_0^\infty a^3 f(a)\, da}{\int_0^\infty a^2 f(a)\, da} \tag{1.3}$$

or the effective diameter $d_{\text{ef}} = 2a_{\text{ef}}$. The radius a_{ef} can be derived from spectral reflectance measurements of snow cover. Therefore, it is often referenced as an optical grain radius.

One must be aware of pitfalls of such a simplified approach for snow optics problems. Indeed, large transparent ice spheres are characterized by rainbow and

a

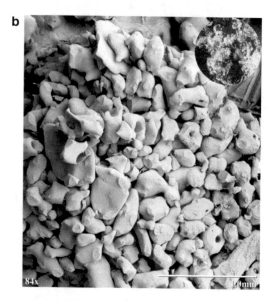

b

Fig. 1.1 a Fresh snow crystals. Courtesy: D. McCoig, **b** Wind-packed snow from Alaska's North slope (USA). Courtesy: Electron and Confocal Microscopy Laboratory, Agricultural Research Service, U. S. Department of Agriculture (https://twistedsifter.com/2013/03/microscopic-images-of-snow-crystals/)

glory effects similar to those observed in water clouds. However, such effects have not been observed for natural snow covers. Therefore, the account for the irregular shape of particles is essential for correct calculations of spectral, angular and polarization characteristics of light reflected and transmitted by snow layers. Meteorological classification of snow crystals has been performed by Magano and Lee (1966). Ishimoto et al. (2018) have derived sizes and shapes of crystals from X-ray computed microtomography imagery. Snow optical properties are modelled using various simplified shapes of ice crystals such as hexagonal plates and columns, Koch fractals, Voronoi crystals, dendrites, etc. The rough crystals are considered as well.

Yet another important parameter of a snowpack is the volumetric concentration c of ice crystals defined as

$$c = \frac{N}{\overline{V}}. \tag{1.4}$$

Here N is the number concentration of ice crystals in unit volume of snow (measured in cm^{-3}) and $\overline{V}$ is the average volume of grains. It follows for spherical particles:

$$\overline{V} = \int_0^\infty v(a) f(a) \, da, \tag{1.5}$$

where $v(a) = 4\pi a^3/3$ is the volume of spherical particles with the radius a. In case of nonspherical particles, one can use the following equation to derive $\overline{V}$:

$$\overline{V} = \int_0^\infty v f(v) \, dv, \tag{1.6}$$

where $f(v)$ is the volume distribution function. The snow density ρ_s is related to the volume concentration of snow grains c:

$$\rho_s = c\rho_i, \tag{1.7}$$

where ρ_i is the density of bulk ice. The value of ρ_i is equal to 0.917 gcm^{-3} at $0\,°$C and standard atmospheric pressure. The snow density is often measured in the filed with typical values in the range 0.1–0.4 gcm^{-3}. The value of volumetric concentration c is often in the range 0.1–0.4 and snow porosity $p = 1 - c$ is in the range 0.6–0.9.

For such high concentrations of ice grains one can observe some order and correlation in the position of ice grains. This in principle may influence the snow optical properties. However, corresponding effects are usually ignored (at least in the visible and near infrared parts of the electromagnetic spectrum). The snow density can be derived weighting a given volume of snowpack. Therefore, the procedure is less involved as compared to the complete characterization of snow microstructure.

1.2 The Snow Specific Surface Area

The specific surface area (SSA) σ of any solid including ice is defined as total surface area S of a material per unit mass M. It follows for snow:

$$\sigma = \frac{S}{M}. \tag{1.8}$$

Let us assume that snow grains are not in direct contact. Then one derives:

$$S = N\overline{S},\ M = N\rho_i \overline{V}. \tag{1.9}$$

Here,

$$\overline{S} = \int_0^\infty s f(s)\, ds \tag{1.10}$$

is the average surface area of ice grains, $f(s)$ is the surface area distribution function. Therefore, it follows:

$$\sigma = \frac{\overline{S}}{\rho_i \overline{V}}. \tag{1.11}$$

If grains touch each other and overlap, the accuracy of equation for the SSA presented above decreases. The theoretical calculation using Eq. (1.11) provides an upper value of SSA as compared to in situ specific snow surface area measurements (say, using the methane adsoption technique). The measurements of SSA can be used to estimate the effective diameter of grains defined as:

$$d_{ef} = \frac{6\overline{V}}{\overline{S}}. \tag{1.12}$$

Namely, it follows:

$$d_{ef} = \frac{6}{\sigma \rho_i}. \tag{1.13}$$

This formula makes it possible to derive the effective grain size for a given snow-pack using snow specific surface area measurements. On the other hand, knowing

Table 1.2 Typical values of SSA for different types of snow

Type of snow	SSA (m^2/kg)
Fresh new snow	50–70
Damp new snow	100–200
Settled snow	200–300
Depth hoar	100–300
Wind packed snow	350–400
Firn (granular)	400–830
Very wet	700–800
Glacier ice	830–917

the effective grain size (say, from optical measurements), one may derive SSA:

$$\sigma = \frac{\zeta}{d_{ef}},\qquad(1.14)$$

where $\zeta = 6/\rho_i$. One shall recall that formulations presented here have larger uncertainties for snow having larger values of snow density due to the presence of many touching and overlapping grains. The typical values of SSA for different types of snow are presented in Table 1.2.

1.3   The Snow Water Equivalent

The snow water equivalent (SWE) is defined as the depth of *water* that would theoretically result if the entire snowpack is melted instantaneously. It can be estimated as follows:

$$SWE = \kappa h.\qquad(1.15)$$

Here h is the depth of the snowpack, $\kappa = \rho_s/\rho_w$, ρ_s is the snow density and $\rho_w = 1$ g/cm^3 is the water density. The SWE is more relevant for hydrological applications as compared to the snow depth because the snow of the same depth can contain different amount of water depending on its density. In particular, it contains less amount of water at the start of winter season as compared to the snow of the same depth at the end of winter season due to the snow densification processes.

In remote sensing applications the value of κ is usually unknown in advance. Therefore, it is parameterized as the function of the snow depth h. In particular, one may use the following approximation (Sturm et al. 2010):

$$\kappa = \kappa_{min} + (\kappa_{max} - \kappa_{min})(1 - e^{-\gamma h}).\qquad(1.16)$$

Table 1.3 The minimal and maximal snow densities and values of the coefficient γ for different snow types (Sturm et al. 2010)

Snow class	κ_{min}	κ_{max}	γ, cm^{-1}
Alpine	0.2237	0.5975	0.0012
Maritime	0.2578	0.5979	0.0010
Prairie	0.2332	0.5941	0.0010
Tundra	0.2425	0.3630	0.0049
Taiga	0.2170	0.2170	0.0000

Here γ is the fitting parameter, κ_{min} and κ_{max} are minimal (at the start of the winter season) and maximal values of the coefficient κ. Typical values of these parameters are given in Table 1.3. More accurate formulation of the dependence $\kappa(h)$ includes the effects of snow aging, wind, and temperature. In particular, snow aging can be accounted for multiplying the exponent in Eq. (1.16) by $\exp(-\Psi t)$, where Ψ is the fitting constant and t is time.

The snow depth, the primary source of spatial variability of snow water equivalent, is an important parameter for many practical applications including various operations in snow-covered terrain and also for hydrological models. It is usually measured by the snow rules over snow boards, if available. The ground—based tripoid lidar systems (terrestrial laser scanner, TLS) are used to monitor the snow depth with a high accuracy (5 cm) in automatic way. The snow depth is found via subtraction of snow—free from snow-covered surface datasets (Deems et al. 2013). Many TLS systems operate at 1550 nm, while most airborne lidar systems work at 503 and 1064 nm. The systems operating at longer wavelengths have generally smaller penetration depths, and, therefore, smaller errors in the snow depth estimation related to laser light scattering inside snow. On the other hand, small reflectivity of snow at near—IR wavelengths limits the operation range. Typical TLS systems deployed to characterize snow and ice are significantly range limited (<150 m) due to very low snow reflectance at 1500 nm. One can use the wavelength 1064 nm to increase the snow reflectivity and operation range. However, TLS systems with a 1064 nm wavelength are not safe at close range using common pulse and scan rates (Deems et al. 2013).

Both snow depth and SWE are important characteristics of snowpack needed for numerous applications. Therefore, they are measured routinely in field and also derived using remote sensing techniques. Sample time series of SWE and snow depth are given in Fig. 1.2.

1.4 Layered Nature and Complex Geometry of Snow Fields

Snow belongs to the class of vertically inhomogeneous turbid media. Its microphysical (e.g., snow grain size, shape, orientation of grains and density) properties and also pollution load change with distance from the snow top. The layered nature of snow is due to occurrence of subsequent snowfalls and possible dust falls and other

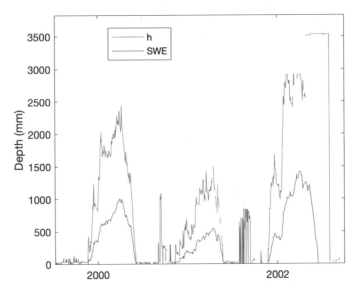

Fig. 1.2 Sample time series (2000–2002) of SWE and snow depth from the Rex River (WA, USA) SNOTEL station. Observations of *h* at times when SWE is zero are spurious (from Hill et al. 2019)

pollution events. The measured snow grain size, shape, density, and temperature profiles for 1.0–1.6 m snowpack are shown in Fig. 1.3. A series of melt/freeze and dust loaded layers and also presence of wetting fronts are clearly seen. Melt/freeze of ice layers are consistently observed beneath dust layers. One can see that density can increase or decrease with snow depth/location in snow layer depending on snowpack. The vertical profile of grain size shows large fluctuations with general tendency of smaller snow crystals to be located closer to the snow top.

In addition, the snow surface is characterized by the presence of sastrugi, snow dunes, etc. Under the action of steady wind, free snow particles can accumulate and drift like the sand grains in barchan dunes. Sastrugi are sharp irregular grooves or ridges, which are formed on a snow surface by wind erosion, saltation and deposition of snow particles. Sastrugi can have a size of several meters and height 20–30 cm (sometimes 1.5 m) with often presence of sharp ice crystals at the top. They are oriented parallel to the prevailing wind direction. Sastrugi are similar to water waves on the surface of ocean. The example of sastrugi is shown in Fig. 1.4, where snow shadowing and brightening effects caused by sastrugi are illustrated. These features are clearly seen in satellite imagery. They influence albedo of snow fields and have important climatic effects.

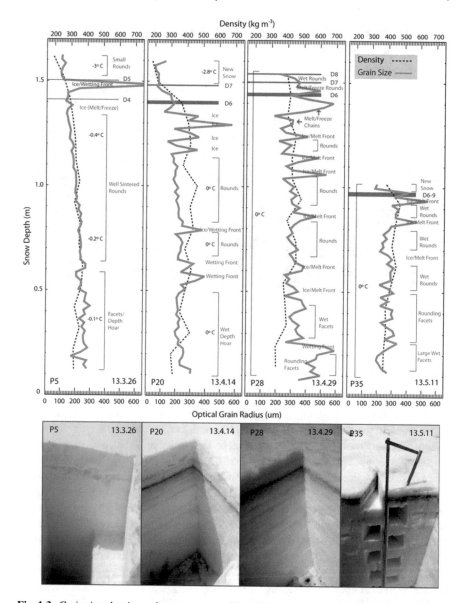

Fig. 1.3 Grain size, density, and temperature profiles with visual stratigraphy identifying prominent features. Pits are labeled with corresponding pit number and date of collection. Pictures from these field days are shown below. The buried polluted snow layers are clearly seen (from Skiles and Painter 2016)

Fig. 1.4 Sastrugi in Antarctica (from NOAA Photo Library—anta0183, Public Domain, https:// commons.wikimedia.org/w/index.php?curid=66437699). Diamond dust above sastrugi and also sparkling crystals on the surface of sastrugi are present

1.5 Snow Impurities: Soot, Dust, and Algae

Natural snow can contain liquid water (wet snow) and various impurities including living matter (see Figs. 1.5, 1.6 and 1.7). Snow absorbs certain polycyclic aromatic hydrocarbons which are organic pollutants known to be toxic and carcinogenic (Nazarenko et al. 2017). The impurity load is characterized by the weight of impurities divided by the weight of a snow sample. The unit ng/g can be substituted by ppbw (parts per billion (10^{-9}) weight). The units μg/g or ppmw (parts per million (10^{-6}) weight) and mg/g or pptw (parts per thousand (10^{-3}) weight) are often used for heavily polluted snow. Di Mauro et al. (2015) reported the values of dust concentration in snow in the range 1–325 μg/g in European Alps after dust deposition events. This leads to reddish snow observed over Alps after dust falls originated in Sahara. Painter et al. (2012) have reported very high (0.2–5 mg/g) annual end-of-melt season dust concentrations at the selected subalpine and alpine sites for time period 2005–2010. The sites are shown in Fig. 1.5. The snow pollution is clearly seen by a naked eye. It is well known that not only dust but also soot present in snow leads to climate forcing via snow albedo (Hansen and Nazarenko 2004; Hadley and Kirchstetter 2012).

The colored (red, green, etc.) algae can present in snow. The examples of snow algae are shown in Figs. 1.6 and 1.7. The characteristic algae absorption peaks are summarized in Table 1.4. Cook et al. (2017a, b) have proposed the physical model, which predicts solar light spectral reflectance and albedo of snowpacks contaminated

a)

b)

Fig. 1.5 a Senator Beck Study Plot (alpine site), 12 May 2009 and **b** Swamp Angel Plot (subalpne site), 13 May 2009. Energy balance/radiation towers are depicted. The total dust loading was 54.6 g/m^3 for year 2009 (from Painter et al. 2012)

Fig. 1.6 Green snow algae (Gray et al. 2020)

with variable concentrations of red snow algae with varying diameters and pigment concentrations. They also estimated the influence of algae on snowmelt.

Fig. 1.7 Red snow algae on Harding Icefield in Alaska (Segawa et al. 2018). Also microscopic view of the dominated snow algae is given on the right upper panel (scale bar is 20 microns). Segawa et al. (2018) have performed single—cell polymerase chain reaction (an isolated cell is shown in right lower panel) for the characterization of collected snow algae

Table 1.4 Spectral absorption features of algal pigments (Cook et al. 2017a)

Pigment	Absorption peaks, nm	Comments
Chlorophyll a	440,680	Two narrow absorption bands
Chlorophyll b	475,650	Two narrow absorption bands
Primary caratenoids	480	One broad band
Secondary carotenoids	460	One broad band
Phycocyanin	610	One broad band (350–700 nm)
Phycoerythrin	450, 525, 575	"Table-shaped" absorption spectrum with a sharp increase in absorption at 420 nm, plateau with a characteristic "triple-peak" morphology to a sharp drop at 580 nm (characteristic for red algae)

It is useful to distinguish two classes of snow surfaces: Case-1 snow (fresh not polluted snow) and Case-2 snow (polluted snow). This is of importance both for the theoretical modelling and for the determination of snow properties from optical measurements. The peculiarities of the spectral reflectance and color of polluted snow is largely determined by various contaminants (see Figs. 1.5–1.7). The reflection of light from clean snow is determined by the ice grain size/shape distributions and optical constants of bulk ice.

1.6 Optical Constants of Ice

The spectral reflectance of polluted snow is determined by the size/shape distributions of ice grains, local optical properties of impurities and complex refractive index $m = n - i\chi$ of the bulk ice. The spectral behavior of optical constants (n, χ) of ice is given in Fig. 1.8. One can see that the value of n slightly decreases with the wavelength. It changes from 1.325 at the wavelength $\lambda = 0.35$ μm to 1.227 at $\lambda = 2.5$ μm. The variation of the imaginary part of ice refractive index χ covers six orders of magnitude. The value of χ is very low in the visible. This leads to the fact that the absorption of light by clean snow in the visible (as far as light reflectance is of concern) can be neglected and snow appears white for human eyes. Such a high reflectivity of snow in the visible is of importance for climate change studies (Hansen and Nazarenko 2004). The snow reflectance is substantially reduced in the infrared region of the electromagnetic spectrum due to larger absorption of light by ice grains in this case.

Water has no absorption bands centred in the visible or near-UV, 0.2–0.7 μm. The observed absorption by water and ice in this region is the tail of near-IR vibrational absorptions, decreasing with decreasing wavelength to extremely small values (Warren 2019). In weakly absorbing regions, the absorption coefficient $\alpha = 4\pi\chi/\lambda$ is measured by transmission (Grenfell and Perovich 1981). Such measurements lead to large uncertainties at small vales of χ at λ below 0.6 μm. The results presented in Fig. 1.8 contain data presented by Warren and Brandt (2008) with modifications for the value of $\chi(\lambda)$ in the spectral range 0.35–0.6 μm proposed by Picard et al. (2016). There is still some uncertainty in the value of $\chi(\lambda)$ below 0.6 μm. The value of $\chi(\lambda)$ is essentially zero in the visible for some purposes. However, small absorption of light by snow in the visible matters for computation of photochemical fluxes in snow. Also there is some impact (below 1%) on the snow albedo in the visible. The imaginary part of ice refractive index is shown in yet another scale in Fig. 1.9. One can see two absorption peaks at 1.5 and 2.0 μm. The horizontal line indicates the spectral region (below ~1.15 μm), where the absorption of light by snow is relatively weak as far as studies of snow local optical properties are of concern. The bands located at 1.5 and 2.0 μm include overlapping overtones and combinations of the fundamental absorption modes. These vibrational modes, because they are intramolecular, are also seen in liquid water and water vapor, shifted somewhat in frequency. The corresponding shifts (also at shorter wavelengths, see Fig. 1.9) can be used to identify the abundance of various thermodynamics states of water in snow (Green et al. 2006).

Ordinary ice is birefringent. Therefore, light refraction depends on the angle between the c-axis of the crystal and the direction of propagation and polarization of

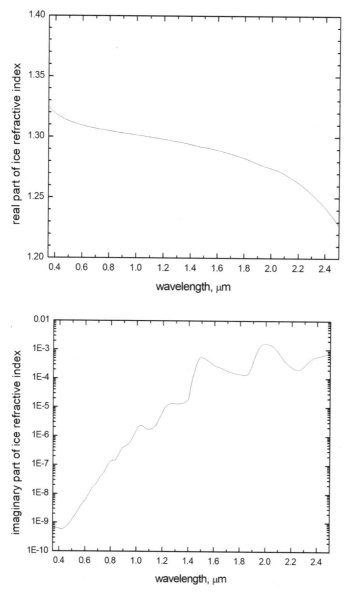

Fig. 1.8 Spectral complex refractive index of ordinary hexagonal ice Ih (Warren and Brand 2008; Picard et al. 2016) at temperature 266 K

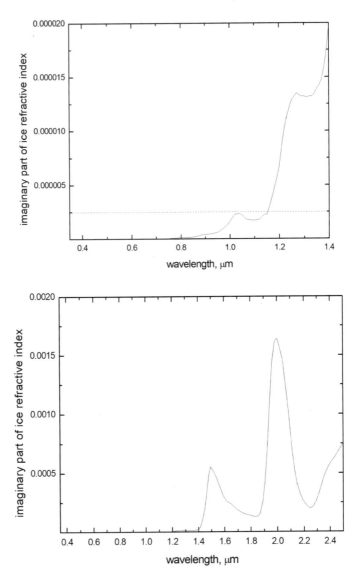

Fig. 1.9 The imaginary part of ice refractive index (Warren and Brand 2008)

incident light. However, the difference in refractive indices of ordinary and extraordinary rays is very small (0.1%) and ice birefringence can be often ignored in snow optics studies. One can neglect the temperature dependence of the complex refractive index of ice in the UV, visible and near IR parts of electromagnetic spectrum. This is not the case in the mid-IR and far-IR bands.

References

Bentley, W.A., and W.J. Humphreys. 1962. *Snow crystals*. New York: Dover.

Cook, J.M., A.J. Hodson, A.J. Taggart, S.H. Mernild, and M. Tranter. 2017a. A predictive model for the spectral "bioalbedo" of snow. *Journal of Geophysical Research Earth Surface* 122: 434–454. https://doi.org/10.1002/2016JF003932.

Cook, J.M., A.J. Hodson, A. Gardner, et al. 2017b. Quantifying bioalbedo: A new physically based model and discussion of empirical methods for characterising biological influence on ice and snow albedo. *The Cryosphere* 11: 2611–2632. https://doi.org/10.5194/tc-11-2611-2017.

Di Mauro, B., B. Fava, L. Ferrero, R. Garzonio, G. Baccolo, B. Delmonte, and R. Colombo. 2015. Mineral dust impact on snow radiative properties in the European Alps combining ground, UAV, and satellite observations. *Journal of Geophysical Research Atmospheres* 120: 6080–6097. https://doi.org/10.1002/2015JD023287.

Deems, J.S., T.H. Painter, and D.C. Finnegan. 2013. Lidar measurement of snow depth: A review. *Journal of Glaciology* 59 (215): 467–479.

Fierz, C., R.L. Armstrong, Y. Durand, P. Etchevers, E. Greene, D. M. McClung, K. Nishimura, P.K. Satyawali, and S. A. Sokratov. 2009: The International Classification for Seasonal Snow on the Ground. IHP-VII Technical Documents in Hydrology N°83, IACS Contribution N°1, UNESCO-IHP, Paris.

Gray, A., M. Krolikowski, P. Fretwell, et al. 2020. Remote sensing reveals Antarctic green snow algae as important terrestrial carbon sink. *Nature Communications* 11: 2527. https://doi.org/10.1038/s41467-020-16018-w.

Green, R.O., T.H. Painter, D.A. Roberts, and J. Dozier. 2006. Measuring the expressed abundance of the three phases of water with an imaging spectrometer over melting snow. *Water Resources Research* 42: W10402. https://doi.org/10.1029/2005WR004509.

Grenfell, T.C., and D.K. Perovich. 1981. Radiation absorption coefficients of polycrystalline ice from 400 to 1400 nm. *Journal of Geophysical Research* 86: 7447–7450. https://doi.org/10.1029/JC086iC08p07447.

Hansen, J., and L. Nazarenko. 2004. Soot climate forcing via snow and ice albedos. *Proceedings of the National Academy of Sciences* 101: 423–428. https://doi.org/10.1073/pnas.2237157100.

Hadley, O., and W. Kirchstetter. 2012. Black-carbon reduction of snow albedo. *Nature Climate Change* 2: 437–440.

Hill, D.F., E.A. Burakowski, R.L. Crumley, J. Keon, J.M. Hu, A.A. Arendt, K.W. Jones, and G.J. Wolken. 2019. Converting snow depth to snow water equivalent using climatological variables. *The Cryosphere* 13: 1767–1784. https://doi.org/10.5194/tc-13-1767-2019.

Ishimoto, H., S. Adachi, S. Yamaguchi, T. Tanikawa, T. Aoki, and K. Masuda. 2018. Snow particles extracted from X-ray computed microtomography imagery and their single scattering properties. *Journal of Quantitative Spectroscopy and Radiative Transfer* 209: 113–128.

Magano, C., and C.V. Lee. 1966. Meteorological classification of natural snow crystals. *Journal of the Faculty of Science, Hokkaido University* 7: 321–362.

Nazarenko, Y., S. Fournier, U. Kurien, et al. 2017. Role of snow in the fate of gaseous and particulate exhaust pollutants from gasoline-powered vehicles. *Environmental Pollution* 223: 665. https://doi.org/10.1016/j.envpol.2017.01.082.

Painter, T., M. Skiles, J. Deems, et al. 2012. Dust radiative forcing in snow of the Upper Colorado River Basin: 1. A 6 year record of energy balance, radiation, and dust concentrations. *Water Resources Research* 48: W07521. https://doi.org/10.1029/2012WR011985.

Picard, G., Q. Libois, and L. Arnaud, L. 2016. Refinement of the ice absorption spectrum in the visible using radiance profile measurements in Antarctic snow. *The Cryosphere* 10: 2655–2672. https://doi.org/10.5194/tc-2016-146.

Segawa, T., R. Matsuzaki, N. Takeuchi, et al. 2018. Bipolar dispersal of red-snow algae. *Nature Communications* 9: 3094. https://doi.org/10.1038/s41467-018-05521-w.

Skiles, S., and T. Painter. 2016. Daily evolution in dust and black carbon content, snow grain size, and snow albedo during snowmelt, Rocky Mountains, Colorado. *Journal of Glaciology* 63 (237): 118–132. https://doi.org/10.1017/jog.2016.125.

Sturm, M., B. Taras, G.E. Liston, C. Derksen, T. Jonas, and J. Lea. 2010. Estimating snow water equivalent using snow depth data and climate classes. *Journal of Hydrometeorology* 11: 1380–1394.

Warren, S.G., and R.E. Brandt. 2008. Optical constants of ice from the ultraviolet to the microwave: A revised compilation. *Journal of Geophysical Research* 113: D14220. https://doi.org/10.1029/2007JD009744.

Warren, S.G. 2019. Optical properties of ice and snow. *Philosophical Transactions, Series a, Mathematical, Physical, and Engineering Sciences* 377 (2146): 20180161. https://doi.org/10.1098/rsta.2018.0161.

Chapter 2
Local Optical Properties of Snowpack

2.1 Geometrical Optics of Large Spherical Particles

2.1.1 Light Scattering

Snow crystals are much larger as compared to the wavelength λ of incident light. Therefore, the interaction of light with a single crystal can be studied in the framework of geometrical optics. To simplify, we consider at first main results of geometrical optics as applied to spherical scatterers with the radius a and the relative complex refractive index $m = n - i\chi$. Let us assume that $x \rightarrow \infty$ and $2x(n - 1) \rightarrow \infty$, where $x = ka$ is the size parameter and $k = \frac{2\pi}{\lambda}$. This makes it possible to ignore phase shifts and possible interference effects.

The interaction of light beam with a sphere with the radius a much larger as compared to the wavelength λ of incident light is demonstrated in Fig. 2.1. The incidence angle i shown in Fig. 2.1 varies from 0 (incidence along the diameter of sphere) to 90° for the grazing incidence. For a given incidence angle, the reflected and refracted beams originate. The reflected beam leaves the particle at the reflection angle equal to the incident angle i. The refracted beam propagates in the direction specified by the refraction angle r as it follows from the Snell's law ($sin\ r = n\ sin\ i$, where it is assumed that $n \ll \chi$ (see Appendix)). One can see that the path of the ray becomes longer as compared to the undisturbed ray propagating in the direction specified by the angle i. The ray deviates in the direction of the center of a particle (focusing effect of the particles). This means that possible impurities (say, soot) located inside of scatterers act as more effective light absorbers (due to larger absorption on longer paths inside particles) as compared to the same impurities outside scatterers. Therefore, external and internal mixture of impurities in snow will lead to different light absorption in snow and also differences in snowmelt processes and timing. The refracted light is partially refracted for the second time and leaves the particle at the angle i. Another portion of the ray is reflected inside the particle and the process is repeating infinite number of times forming a complex scattering pattern around the particle. In case of absorbing particles, one needs to

© Springer Nature Switzerland AG 2021
A. Kokhanovsky, *Snow Optics*,
https://doi.org/10.1007/978-3-030-86589-4_2

Fig. 2.1 Ray tracing in a
sphere
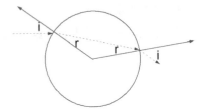

account for the absorption processes on the path l of the ray inside the particle. This is done taking into account the exponential attenuation of the ray intensity J on the distance L according to the following law: J $\exp(-\alpha L)$, where $\alpha = \frac{4\pi\chi}{\lambda}$ is the bulk ice absorption coefficient, χ is the imaginary part of ice refractive index at the wavelength λ.

Let us consider first the reflected light beam and introduce the complementary angles

$$\tau = \frac{\pi}{2} - i, \tau' = \frac{\pi}{2} - r. \tag{2.1}$$

The use of these angles makes the final equations somewhat simpler. In particular, it follows from the Snell's law at $\chi \ll n$:

$$cos\tau' = \frac{cos\tau}{n}. \tag{2.2}$$

This also means that

$$sin\tau' = \sqrt{1 - \left[\frac{cos\tau}{n}\right]^2}. \tag{2.3}$$

A finite pencil of light is characterized by the spread in the azimuthal angle $d\varphi$ and also $d\tau$. The flux of energy in this pencil is $I_0 a^2 cos\tau sin\tau d\tau d\varphi$ for a light coming in the direction specified by the angle τ (Shifrin, 1951; van de Hulst, 1957). Here I_0 is the intensity of incident light. The reflected part of the energy is $I_0 R_j a^2 cos\tau sin\tau d\tau d\varphi$, for the light polarized perpendicular (j = 1) or perpendicular (j = 2) to the incidence plane. It follows (Fresnel 1831; van de Hulst 1957):

$$R_1 = \left|\frac{sin\tau - msin\tau'}{sin\tau + msin\tau'}\right|^2, R_2 = \left|\frac{msin\tau - sin\tau'}{msin\tau + sin\tau'}\right|^2. \tag{2.4}$$

It is known that $\chi \ll n$ for ice in the visible and near infrared. Therefore, one can assume that $m = n$ in Eq. (2.4).

The emergent pencil of light spreads into a solid angle $d\Omega = \sin\theta d\theta d\varphi$. The intensity of reflected light beam at a large distance r in the direction specified by the scattering angle θ is found dividing reflected light flux by the area $dA = r^2 d\Omega$

Namely it follows:

$$I_{r,j} = \frac{I_0 R_j a^2 \cos\tau \sin\tau \, d\tau \, d\varphi}{r^2 \sin\theta \, d\theta \, d\varphi} = I_0 R_j D(a/r)^2, \tag{2.5}$$

where

$$D = \frac{\sin 2\tau}{2 \upsilon \sin\theta} \tag{2.6}$$

and

$$\upsilon = \frac{d\theta}{d\tau} = \left| \frac{d\theta'}{d\tau} \right|. \tag{2.7}$$

Here, the angle

$$\theta' = 2\tau \tag{2.8}$$

describes the deviation of the reflected ray from the initial direction. It defines the scattering angle in the range (0–180°):

$$\theta = 2\pi K + q\theta', \tag{2.9}$$

where K is an integer and $q = -1$ or $q = 1$. It is easy to show that for other beams (two times refracted, refracted-reflected-refracted, etc.), it follows:

$$I_{s,j} = I_0 (1 - R_j)^2 R_j^{s-1} D(a/r)^2, \quad s = 1, 2, 3, \ldots \tag{2.10}$$

In case of *absorbing* spheres and $\chi \ll n$, it follows:

$$I_{s,j} = I_0 (1 - R_j)^2 R_j^{s-1} D(a/r)^2 \exp(-\alpha l), \tag{2.11}$$

where $l = z \sin(\tau')$, $\alpha = 2k\chi$, $k = \frac{2\pi}{\lambda}$, $z = 2as$, $s = 1, 2, 3, \ldots$.
Summing up, the total geometrical optics scattered intensity can be written as

$$I^g = \frac{I_0}{2k^2 r^2} (x^2 (R_1 + R_2) D + \sum_{s=1}^{\infty} (i_{1s} + i_{2s})), \tag{2.12}$$

where

$$i_{js} = x^2 \varepsilon_j^s D, \tag{2.13}$$

$$\varepsilon_j^s = \left(1 - R_j\right)^2 R_j^{s-1} \exp\left(-cs\sin\tau'\right), \tag{2.14}$$

$$x = ka, c = 2\alpha a. \tag{2.15}$$

The geometrical optics scattering cross section is defined as an integral of intensity given by Eq. (2.12). Namely it follows:

$$C_{sca}^g = \frac{1}{I_0} \int_0^{2\pi} d\varphi \int_0^{\pi} I^g(\theta) \ r^2 \sin\theta d\theta. \tag{2.16}$$

It follows from Eqs. (2.16) and (2.12):

$$C_{sca}^g = \frac{\pi a^2}{2} \sum_{j=1}^{2} \int_0^{\frac{\pi}{2}} \left(R_j + P_j\right) \sin 2\tau d\tau, \tag{2.17}$$

where

$$P_j = \sum_{s=1}^{\infty} \left(1 - R_j\right)^2 R_j^{s-1} \exp\left(-cs \ \sin\tau'\right). \tag{2.18}$$

We derive performing the summation in Eq. (2.18):

$$P_j = \frac{\left(1 - R_j\right)^2 e^{-cs\sin\tau'}}{1 - R_j e^{-cs\sin\tau'}}. \tag{2.19}$$

Therefore, it follows (Kokhanovsky and Zege 1995):

$$C_{sca}^g = \frac{\pi a^2}{2} \sum_{j=1}^{2} \int_0^{\frac{\pi}{2}} \left(R_j + \frac{\left(1 - R_j\right)^2 e^{-cs\sin\tau'}}{1 - R_j e^{-cs\sin\tau'}}\right) \sin 2\tau d\tau. \tag{2.20}$$

One can see that the geometrical optics scattering cross section of an absorbing spherical particle can be expressed via a single integral over the incidence angle τ.

As a matter of fact, scattered light around large spherical particle can be presented as a sum of contribution due to two physical processes: diffraction and reflection/refraction of rays at the surface of particles. The cross section shown in Eq. (2.21) accounts just for the reflection and refraction processes. The intensity of diffracted light can be presented by the following equation (van de Hulst, 1957):

$$I_{sca}^d = \frac{I_0 a^2 J_1^2(\theta x)}{\theta^2 r^2},$$

(2.21)

where $J_1(\theta x)$ is the Bessel function. It follows from Eqs. (2.16) and (2.21):

$$C_{sca}^d = 2\pi a^2 \int_0^\infty J_1^2(y) y^{-1} dy,$$

(2.22)

where $y = \theta x$ and we have assumed that $x \to \infty$ and $sin\theta \approx \theta$ at small scattering angles. We extended the upper limit integration to infinity taking into account that the scattering angle is fixed and small but $x \to \infty$. The integral in Eq. (2.22) can be evaluated analytically:

$$2 \int_0^\infty J_1^2(y) y^{-1} dy = 1.$$

(2.23)

Then it follows:

$$C_{sca}^d = \pi a^2.$$

(2.24)

Therefore, total scattering cross section is given by the following expression:

$$C_{sca} = \pi a^2 (1 + W),$$

(2.25)

where

$$W = \frac{1}{2} \int_0^{\frac{\pi}{2}} (w_1 + w_2) sin2\tau d\tau,$$

(2.26)

and

$$w_j = R_j + \frac{(1 - R_j)^2 e^{-csin\tau'}}{1 - R_j e^{-csin\tau'}}.$$

(2.27)

Let us introduce the scattering efficiency factor:

$$Q_{sca} = \frac{C_{sca}}{\pi a^2}.$$

(2.28)

It follows for this factor:

$$Q_{sca} = 1 + \frac{1}{2} \int\limits_{0}^{\frac{\pi}{2}} (w_1 + w_2) sin2\tau d\tau. \tag{2.29}$$

The scattering efficiency factor can be also derived using Maxwell electromagnetic theory (Maxwell 1873; Mie 1908). The derivations are much more involved as compared to our discussion valid for the case of large spherical scatterers. The intercomparison of geometrical optics and Mie theory calculations of Q_{sca} for monodispersed ice spheres with radii 100, 500, and 1000 microns in the spectral range 0.3–2.5 microns is given in Fig. 2.2. As one might expect, it follows that the accuracy of geometrical optics approximation increases with the size parameter of particles. The largest deviations occur for 100 μm spheres (fine grained snow). The high frequency oscillations in Fig. 2.2 are due to interference of various geometrical optics rays (van de Hulst, 1957) ignored in the theory described above. They do not appear if one takes into account the polydispersity of snow grains.

One can derive from Eq. (2.27) at $c = 0$ (nonabsorbing particles): $w_j = 1$ and, therefore, $W = 1$ (see Eq. 2.26). This means that

$$Q_{sca} = 2 \tag{2.30}$$

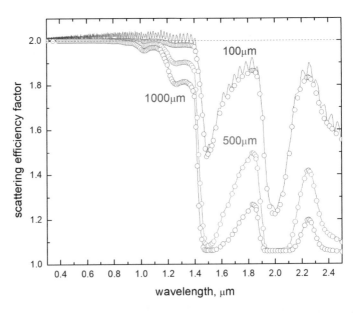

Fig. 2.2 The intercomparison of geometrical optics (symbols) and Mie theory (lines) calculations of Q_{sca} for monodispersed ice spheres with radii 100, 500, and 1000 microns

in this case. This result is very important. It shows that the scattering efficiency factor does not depend on the refractive index of particles and their size in the geometrical optics limit for large nonabsorbing spheres. Equation (2.30) is valid for nonspherical particles as well (van de Hulst 1957).

Let us use the following approximation valid at small values of the attenuation parameter c:

$$w_j = 1 - c sin\tau'. \tag{2.31}$$

Therefore, it follows:

$$Q_{sca} = 2 - Ac, \tag{2.32}$$

where the integral

$$A = \int_0^{\frac{\pi}{2}} sin\tau' sin2\tau d\tau \tag{2.33}$$

can be evaluated analytically using Eq. (2.3) and the substitution: $u = cos^2\tau, du = sin2\tau d\tau$. The answer is:

$$A = \frac{2}{3}n^2(1 - (1 - n^{-2})^{\frac{3}{2}}). \tag{2.34}$$

One can see that $A \to \frac{2}{3}$ as $n \to 1$. Equation (2.29) can be parameterized as follows:

$$Q_{sca} = 1 + \rho + (1 - \rho)\exp(-vc), \tag{2.35}$$

where v is the fitting parameter and the integral

$$\rho = \frac{1}{2}\int_0^{\frac{\pi}{2}} (R_1 + R_2)sin2\tau d\tau \tag{2.36}$$

can be evaluated analytically at $\chi \ll n$. Namely, it follows:

$$\rho = \frac{8n^4(1 + n^4)\ln n}{(1 + n^2)(1 - n^4)^2} + \frac{n^2(1 - n^2)\ln[\frac{n-1}{n+1}]}{(1 + n^2)^3} - \frac{\sum_{j=0}^7 p_j n^j}{3(1 + n)(1 + n^2)(1 - n^4)}, \tag{2.37}$$

where $p_j = (-1, -1, -3, 7, -9, -13, -7, 3)$.

The fitting parameter ν can be found from the condition that Eq. (2.35) must coincide with Eq. (2.32) as $c \to 0$. It provides the correct behavior of C_{sca} at small values of c. Then it follows:

$$\nu = \frac{A}{1 - \rho}, \tag{2.38}$$

where A is given by Eq. (2.34). The value of ρ given by Eq. (2.37) can be parameterized as follows at n = 1.2–1.4:

$$\rho = 0.0123 + 0.1622(n - 1). \tag{2.39}$$

The intercomparison of calculations according to Eqs. (2.29) and (2.35) is shown in Fig. 2.3. It follows that analytical Eq. (2.35) can be used with the error less than 1% at $n = 1.31$. One concludes from Eq. (2.29) and Fig. 2.3 that the scattering efficiency factor for large spherical scatters changes from 2 for nonabsorbing particles to $1 + \rho$ for strongly absorbing particles. The value of ρ depends on the refractive index n and close to 0.06 at $n = 1.31$. The same is true for large nonspherical particles.

The normalized scattering light intensity

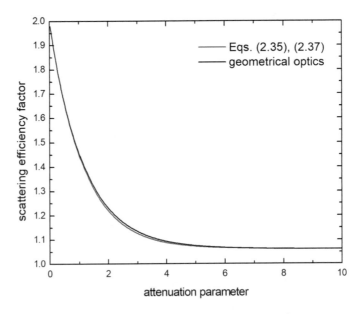

Fig. 2.3 The dependence of the scattering efficiency factor on the attenuation parameter c calculated using Eqs. (2.35), (2.37) (red line) and geometrical optics Eq. (2.29) (black line) at $n = 1.31$

$$p^g(\theta) = \frac{4\pi(I_1 + I_2)}{k^2 C_{sca}^g} \tag{2.40}$$

is called phase function. It follows:

$$I_j = \sum_{s=0}^{\infty} i_{js} \tag{2.41}$$

and i_{js} for each ray is given by Eq. (2.13). Details of numerical calculations of the phase function for large spherical particles are given by Shifrin (1951) and van de Hulst (1957). Zhou et al (2003) provide the code, which can be used for the calculations of phase functions of large spherical particles in the framework of geometrical optics.

The total phase function, which accounts for both geometrical optics and diffraction components, can be presented in the following way:

$$p(\theta) = \frac{C_{sca}^d p^d(\theta) + C_{sca}^g p^g(\theta)}{C_{sca}^d + C_{sca}^g}, \tag{2.42}$$

where

$$p^d(\theta) = \frac{x^2 F(\theta)}{2} \tag{2.43}$$

and

$$F(\theta) = \frac{4 J_1^2(\theta x)}{x^2 \theta^2}. \tag{2.44}$$

The phase function $p(\theta)$ is normalized as follows:

$$\frac{1}{2} \int_0^\pi p(\theta) \sin\theta d\theta = 1. \tag{2.45}$$

Similar equations hold for $p^{d,g}(\theta)$. The phase function gives the conditional probability of light scattering in a given direction and can be used to calculate various statistical parameters relevant to the propagation of light in turbid media. In particular, the average cosine of scattering angle g can be derived:

$$g = \frac{1}{2} \int_0^\pi p(\theta) \sin\theta \cos\theta d\theta. \tag{2.46}$$

Let us derive useful approximation for the average cosine of scattering angle in the framework of geometrical optics. Then it follows (van de Hulst 1957; Kokhanovsky and Zege 1995):

$$g = \frac{1 + W g^G}{1 + W},$$ (2.47)

where

$$g^G = \frac{1}{2W} \int_0^{\frac{\pi}{2}} (\epsilon_1 + \epsilon_2) \sin 2\tau d\tau,$$ (2.48)

$$\epsilon_j = \frac{\left(1 - R_j\right)^2 \cos 2(\tau - \tau') e^{-c\xi} + R_j \cos 2\tau \left(1 - e^{-2c\xi}\right)}{+ 2R_j^2 \cos 2\tau e^{-c\xi} \left(e^{-c\xi} - \cos 2\tau'\right)}}{1 - 2R_j e^{-c\xi} \cos 2\tau' + R_j^2 e^{-c\xi}},$$

$$\xi = \sqrt{1 - \frac{\cos^2 \tau}{n^2}}$$ (2.49)

The inter-comparison of asymmetry parameters calculated using Eqs. (2.47–2.49) and Mie theory is shown in Fig. 2.4. One can see that size dependence of the

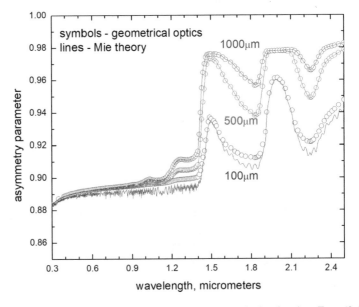

Fig. 2.4 The intercomparison of asymmetry parameters calculated using Eqs. (2.47–2.49) (symbols) and Mie theory (lines) for the radii of particles equal to 100, 500, and 1000 μm

asymmetry parameter is weak in the visible. The results of geometrical optics and Mie calculations almost coincide for large values of size parameter. Therefore, fast geometrical optics calculations can be used instead of tedious Mie computations for large scatterers.

It follows at $c = 0$:

$$\in_j (0) = \frac{(1 - R_j)^2 cos2(\tau - \tau') + 2R_j^2 cos2\tau(1 - cos2\tau')}{1 - 2R_j cos2\tau' + R_j^2} \tag{2.50}$$

and

$$g(0) = \frac{1 + g_0^G}{2}, \tag{2.51}$$

where

$$g_0^G = \frac{1}{2} \int_0^{\frac{\pi}{2}} (\in_1 (0) + \in_2 (0)) \sin 2\tau d\tau \tag{2.52}$$

In the opposite case of strongly absorbing particles ($c \to \infty$) one derives:

$$\in_j(c \to \infty) = R_j cos2\tau \tag{2.53}$$

and

$$g_\infty \equiv g(c \to \infty) = \frac{1 + K}{1 + \rho}, \tag{2.54}$$

where ρ is given by Eq. (2.37) and

$$K = \frac{1}{4} \int_0^{\frac{\pi}{2}} (R_1 + R_2)sin4\tau d\tau. \tag{2.55}$$

This integral can be evaluated analytically. The answer is

$$K = \frac{8n^4(n^6 - 3n^4 + n^2 - 1) \ln n}{(1 + n^2)^2 (1 - n^4)^2}$$
$$+ \frac{(1 - n^2)^2 (n^8 + 12n^6 + 54n^4 - 4n^2 + 1) \ln\left[\frac{n-1}{n+1}\right]}{16(1 + n^2)^4}$$

$$+ \frac{\sum_{j=1}^{12} q_j n^j}{24(1+n)(1+n^2)^2(n^4-1)}, \tag{2.56}$$

where $q_j = (-3, 13, -89, 151, 186, 138, -282, 22, 25, 25, 3, 3)$.

Unfortunately, analytical integration for the asymmetry parameter g at any c is not possible. However, one can use the following uniform approximation:

$$g = g_\infty - (g_\infty - g_0)\exp(-c\eta), \tag{2.57}$$

where $g_\infty = g(\infty)$, $g_0 = g(0)$ and the value of η can be estimated, e.g., using the asymptotic behavior of g given by Eq. (2.46) at small values of c. Then it follows:

$$g = g_0 + cX, \tag{2.58}$$

where

$$X = P + g_0 \Upsilon, \tag{2.59}$$

$$P = \frac{1}{2} \int_0^{\frac{\pi}{2}} D_j(\tau) sin2\tau d\tau, \quad \Upsilon = n^2(1-(1-n^{-2})^{\frac{3}{2}})/3, \tag{2.60}$$

$$D_j(\tau) = \left[\frac{Z_j}{N_j} + \frac{M_j L_j}{N_j^2}\right]\sqrt{1 - \frac{cos^2\tau}{n^2}}. \tag{2.61}$$

Here, $Z_j = 2R_j(1-2R_j)cos2\tau - (1-R_j^2)cos2\tilde{\tau} + 2R_j^2 cos2\tau cos2\tau'$, $L_j = R_j^2 - 2R_j cos2\tau'$, $\tilde{\tau} = \tau - \tau'$, $M_j = (1-R_j)^2 cos2\tilde{\tau} + 2R_j^2(1-cos2\tau')cos2\tau$, $N_j = 1 - 2R_j cos2\tau' + R_j^2$. It follows from Eq. (2.57) at small values of c:

$$g = g_0 + c\eta(g_\infty - g_0). \tag{2.62}$$

Therefore, one derives from Eqs. (2.58) and (2.62):

$$\eta = \frac{X}{g_\infty - g_0}, \tag{2.63}$$

where X is defined by Eq. (2.59). The parameters g_0, g_∞ and η depend only on the real part of ice refractive index at $\chi \ll n$. They can be parameterized as follows:

$$g_0 = 1.006 - 0.3641(n-1), \quad g_\infty = 1.008 - 0.11(n-1),$$

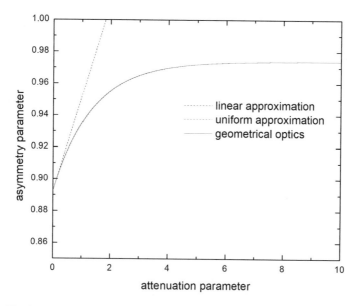

Fig. 2.5 The dependence of the asymmetry parameter of ice spheres on the attenuation parameter at the real part of ice refractive index 1.31 and radius of particles equal to 1000 μm. The wavelength is equal to 1 μm. The results of calculations using the geometrical optics approach (Eq. 2.47), uniform approximation (Eq. 2.57), and linear approximation (Eq. 2.62) are shown

$$\eta = 0.3639 + 1.676(n-1) - 1.6284(n-1)^2 \tag{2.64}$$

at $n = 1.2$–1.4. The accuracy of Eq. (2.57) is presented in Fig. 2.5. The error is below 0.5%. The accuracy of linear approximation (see Eq. (2.62)) is also shown. It can be used for the attenuation parameters c smaller than 0.5.

2.1.2 Light Absorption and Extinction

The intensity of transmitted light for the j-*th* polarization interacted s-time with the surface of the particle is given by Eq. (2.11). In particular, it follows for the transmitted beam at $s = 1$:

$$I_j = I_0(1-R_j)^2 D(a/r)^2 \exp(-csin\tau'). \tag{2.65}$$

It follows in the absence of absorption:

$$I_j = I_0(1-R_j)^2 D(a/r)^2. \tag{2.66}$$

Therefore, the absorbed light intensity on the ray path can be presented as

$$I_j^{abs}(s = 1) = I_0(1 - R_j)^2 D(a/r)^2\left(1 - \exp(-csin\tau')\right) \tag{2.67}$$

at $s = 1$. It follows at $s = 2$:

$$I_j^{abs}(s = 2) = I_0(1 - R_j)^2 D(a/r)^2)R_j\left(1 - \exp(-2csin\tau')\right). \tag{2.68}$$

One derives summing all rays:

$$I_j^{abs} = \frac{I_0(1 - R_j)Da^2(1 - \exp(-csin\tau'))}{(1 - R_j\exp(-csin\tau'))r^2}. \tag{2.69}$$

The absorption cross section is defined as

$$C_{abs} = \frac{1}{I_0}\int\limits_{0}^{2\pi} d\varphi \int\limits_{0}^{\pi} (I_r^{abs}(\theta) + I_p^{abs}(\theta))r^2 sin\theta d\theta. \tag{2.70}$$

Therefore, one derives:

$$C_{abs} = M\pi a^2, \tag{2.71}$$

where

$$M = \frac{1}{2}\int\limits_{0}^{\frac{\pi}{2}} (m_1 + m_2)sin2\tau d\tau, \quad m_j = \frac{(1 - R_j)(1 - e^{-csin\tau'})}{1 - R_je^{-csin\tau'}}. \tag{2.72}$$

The value of M coincides with the absorption efficiency factor

$$Q_{abs} = \frac{C_{abs}}{\pi a^2}. \tag{2.73}$$

It is easy to show that

$$M + W = 1. \tag{2.74}$$

This means that both absorption and scattering cross section are determined by a single geometrical optics integral (e.g., W, see Eq. 2.26). Therefore, we can write:

$$C_{abs} = \pi a^2(1 - W), \ C_{sca} = \pi a^2(1 + W). \tag{2.75}$$

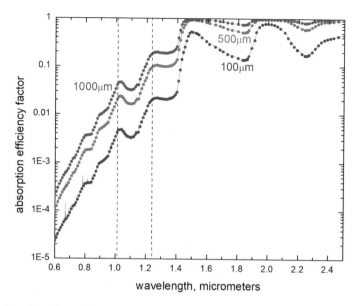

Fig. 2.6 The dependence of the absorption efficiency factor of ice spheres of various radii on the wavelength in the near infrared (solid lines–Mie theory, symbols–geometrical optics)

It follows for the efficiency factors:

$$Q_{abs} = 1 - W, \; Q_{sca} = 1 + W. \tag{2.76}$$

An important point is that efficiency factors depend on just two parameters in the geometrical optics approximation (at $\chi \ll n$). They are real part of refractive index n and attenuation parameter $c = \alpha d$, which is equal to the product of bulk ice absorption coefficient and the diameter of particles $d = 2a$. The accuracy of Eq. (2.76) for the absorption efficiency factor is shown in Fig. 2.6.

The extinction cross section C_{ext} is defined as the sum of scattering C_{sca} and absorption C_{abs} cross sections:

$$C_{ext} = C_{sca} + C_{abs}. \tag{2.77}$$

Therefore, we arrive to the following equation for the extinction cross section :

$$C_{ext} = 2\pi a^2. \tag{2.78}$$

This equation is valid at any c in the framework of the geometrical optics approximation ($x \to \infty, 2x(n-1) \to \infty$). This result is very important. It shows that the extinction cross section of large spherical particles does not depend neither on the

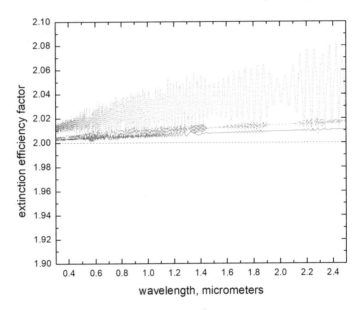

Fig. 2.7 The spectral dependence of the extinction efficiency factor of ice spheres derived using Mie theory at radii equal to 25 (blue line), 50 (green line), and 100 (red line) microns

wavelength nor on the refractive index of particles. The extinction efficiency factor $Q_{ext} = \frac{C_{ext}}{\pi a^2}$ is equal to 2 for large spherical particles independently of their actual size and chemical composition as far as $x \to \infty$ and $2x(n-1) \to \infty$, where x is the size parameter. The spectral dependence of the extinction efficiency factor derived using Mie theory for monodispersed ice spheres with radii 25, 50, and 100 microns is shown in Fig. 2.7. The oscillating feature is due to the interference of waves passing the spherical scatterer of a given radius. The spectral oscillations disappear for realistic polydispersed media such as snow. It follows that the error of the asymptotic approximation ($Q_{ext} = 2$) is better than 1% at the spectral range 0.3–2.5 μm and radii larger than 100 microns typical for snow covers (see Table 1.1).

Let us assume that particles are weakly absorbing and, therefore, $c \to 0$. Then it follows from Eqs. (2.77) and (2.32):

$$C_{abs} = B\alpha V, \tag{2.79}$$

where $B = 3A/2$ or

$$B = n^2(1 - (1 - n^{-2})^{\frac{3}{2}}). \tag{2.80}$$

We can see that $B \to 1$ as $n \to 1$ (optically soft particles). It follows from Eq. (2.79) that absorption cross section of large spherical particles is proportional to the volume of particles V and the bulk absorption coefficient α. This remains true for nonspherical particles as well. However, the coefficient B differs for particles of different shapes. It follows for arbitrarily shaped optically soft large particles:

$$C_{abs} = \alpha V. \tag{2.81}$$

The useful approximation for the value of C_{abs} valid at any values of n and c can be derived from Eqs. (2.35) and (2.77):

$$C_{abs} = \pi a^2 (1 - \rho)(1 - \exp(-vc)). \tag{2.82}$$

The parameter v can be found from the condition that C_{abs} must coincide with Eq. (2.75) at small values of c. Then it follows:

$$v = \frac{2B}{3(1 - \rho)}. \tag{2.83}$$

The dependence of the absorption efficiency factor

$$Q_{abs} = \frac{C_{abs}}{\pi a^2} \tag{2.84}$$

on the wavelength for several sizes of spherical particles is shown in Fig. 2.8a. It is seen that errors of geometrical optics approximation decrease with the radius of particles a being negligible for typical (see Table 1.1) snow grain diameters. It follows from Eqs. (2.82) and (2.84) for the absorption efficiency factor:

$$Q_{abs} = (1 - \rho)(1 - \exp(-vc)), \tag{2.85}$$

where the parameter ρ can be parametrized as follows:

$$\rho = 0.0123 + 0.1622(n - 1) \tag{2.86}$$

at n in the range 1.2–1.4. It follows for the scattering efficiency factor in the same uniform approximation:

$$Q_{sca} = 1 + \rho + (1 - \rho) \exp(-vc). \tag{2.87}$$

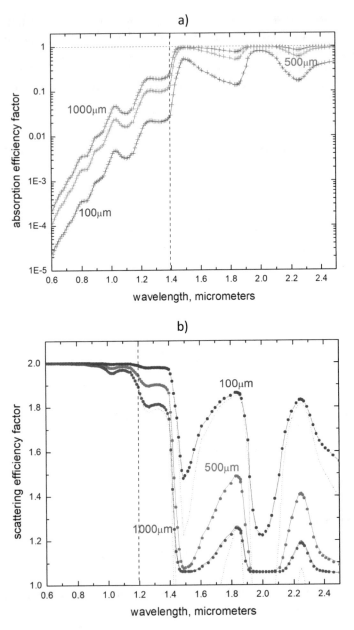

Fig. 2.8 a The absorption efficiency factor of ice spheres with radii 100, 500, and 1000 μm; **b** The scattering efficiency factor of ice spheres with radii 100, 500, and 1000 μm (solid line—geometrical optics, dash line—linear approximation, symbols—uniform approximation). The linear approximation can be used at the wavelengths smaller than 1.2 μm

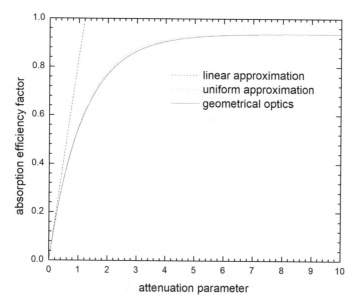

Fig. 2.9 The dependence of the absorption efficiency factor of ice spheres on the attenuation parameter c at the real part of ice refractive index 1.31 and radius of particles equal to 1000 calculated using geometrical optics approximation and also linear and uniform approximations. The wavelength is equal to 1 μm. Similar accuracy of approximations holds for the scattering efficiency factor ($Q_{sca} = 2 - Q_{abs}$)

The accuracy of approximations for the values of Q_{abs} and Q_{sca} is given in Fig. 2.8. It follows that the error of the uniform approximation is better than 2%. One concludes from Eq. (2.85) and Fig. 2.9 that the absorption efficiency factor for large spherical scatters changes from 0 for nonabsorbing particles to $1-\rho$ for strongly absorbing particles. The same is true for large nonspherical particles. The value of ρ is given by Eq. (2.37) at $\chi \ll n$, which is valid assumption for ice in the visible and near-infrared regions of electromagnetic spectrum. The absorption efficiency factor Q_{abs} is directly proportional to the attenuation parameter c for weakly absorbing large spherical particles: $Q_{abs} = Ac$, where the coefficient A depends on the refractive index of particles (see Eq. 2.34).

The parameters introduced above can be used to calculate other light scattering characteristics useful in radiative transfer studies. In particular, the results for the single scattering albedo $\omega_0 = \frac{Q_{sca}}{Q_{ext}}$ and the probability of photon absorption (PPA) $\beta = 1 - \omega_0$ of monodispersed spherical particles are shown in Fig. 2.10a, b.

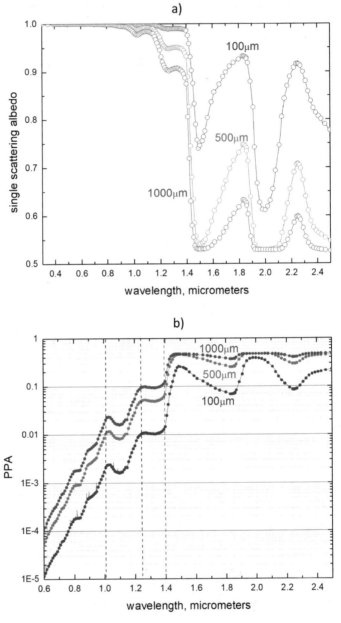

Fig. 2.10 The spectral dependence of single scattering albedo **a** and probability of photon absorption **b** for monodispersed ice spheres with radii 100, 500, and 1000 μm (symbols- geometrical optics, lines–Mie theory)

2.2 Local Optical Properties of Snowpack

2.2.1 Integral Light Scattering and Absorption Characteristics of Large Nonspherical Scatterers

Particles in snow are much larger as compared to the wavelength of the incident light. Also they have nonspherical shapes and respective phase shifts are large. Therefore, the extinction cross section can be presented in the following way (van de Hulst 1957):

$$C_{ext} = 2\Sigma, \tag{2.88}$$

where Σ is the cross section of the particle in the direction perpendicular to the incident light. It follows that extinction cross section of a large nonspherical particle coincides with that of a sphere with the radius $a = \sqrt{\frac{\Sigma}{\pi}}$.

Let us assume that nonspherical particles are convex and randomly oriented. Then it follows for the average value of Σ:

$$\langle \Sigma \rangle = \frac{\overline{\Phi}}{4}, \tag{2.89}$$

where $\overline{\Phi}$ is the average surface area of the particles. Therefore, it follows: $C_{ext} = \frac{\overline{\Phi}}{2}$. The phase function, asymmetry parameter, scattering cross section and absorption cross section of nonspherical particles larger as compared to the wavelength of the incident light can be calculated using the ray tracing approach. This has been done for various shapes of particles and corresponding databases are available online.

The simple geometrical optics approximation for the average absorption cross section of randomly oriented nonspherical particles of the same size can be presented in the form similar to that for spherical particles:

$$C_{abs} = (1 - \rho)(1 - \exp(-wc))\langle \Sigma \rangle, \tag{2.90}$$

where $c = \alpha D$, $D = \frac{3V}{2\langle \Sigma \rangle}$, $\langle \Sigma \rangle$ is given by Eq. (2.89), V is the volume of particles, and w is the fitting parameter. One can see that C_{abs} for large strongly absorbing nonspherical partilces in random orientation can be presented

$$C_{abs} = (1 - \rho)\langle \Sigma \rangle. \tag{2.91}$$

It follows in the case of weakly absorbing particles:

$$C_{abs} = B\alpha V, \tag{2.92}$$

where

$$B = \frac{3}{2}(1 - \rho)w. \tag{2.93}$$

Clearly, one derives: $B \rightarrow 1$, $\rho \rightarrow 0$ as $n \rightarrow 1$ for particles of arbitrary shapes. This means that $w \rightarrow \frac{2}{3}$ as $n \rightarrow 1$. It is of importance that parameter B does not depend on the size of particles for large weakly absorbing scatterers. The dependence of B on the shape of particles has been studied by Libois et al. (2013). The computations have been shown that the value of B varies in the range 1.25 (spheres) to 2.09 (spheroids with aspect ratio 0.5). The following variety of shapes have been considered: cylinders, spheres, spheroids, hexagonal plates, fractal particles, and cuboids. The analysis of measurements performed by Libois et al. (2013) have shown that the most frequent value of B is in the range 1.6–1.9 for natural snowpacks. Taking $B = 7/4$ as a representative value, we derive from Eq. (2.93) at $n = 1.31$: $w = 1.24$. This constant can be used in conjunction with Eq. (2.90) to study the dependence of the absorption cross section of snow grains on the wavelength (under assumption that one can neglect the spectral variation of the real part of the ice refractive in the spectral range under study).

The asymmetry parameter for large nonspherical partilces of the same size and shape in random orientation can be presented in the way similar to the case of spherical scatterers:

$$g = g_\infty - (g_\infty - g_0) \exp(-\gamma c), \tag{2.94}$$

where γ is the fitting parameter, $c = \alpha D$. It follows that $g = g_\infty$ at $c \rightarrow \infty$. The value of g_∞ is given by Eq. (2.54) in case of randomly oriented nonspherical convex particles. The values of g_0 and γ depend on the shape of particles.

Equations given above are valid for the convex randomly oriented particles of a given shape and size. In reality particles of different shapes and sizes present in snow. Therefore, Eqs. (2.90), (2.94) must be averaged with respect to the size and shape of particles. In particular, it follows for the averaged quantities:

$$\langle C_{ext} \rangle = 2\langle \Sigma \rangle, \tag{2.95}$$

$$< C_{abs} >= (1 - \rho)(1 - \exp(-\alpha p \overline{D}))\langle \Sigma \rangle, \tag{2.96}$$

$$< g >= g_\infty - (g_\infty - \langle g_0 \rangle) \exp\left(-\alpha q \overline{D}\right), \tag{2.97}$$

where $\langle \Sigma \rangle$ is given by Eq. (2.89) for convex particles in random orientation and

$$\overline{D} = \frac{6\overline{V}}{\Phi} \tag{2.98}$$

is the effective grain diameter. Clearly, one derives: $\langle C_{sca} \rangle = \langle C_{ext} \rangle - \langle C_{abs} \rangle$. One can assume that $\langle g_0 \rangle$ is equal to 0.76 as measured in situ for ice clouds in the visible (Garret et al. 2001). Generally, the value of $\langle g_0 \rangle$ increases with the roundness of grains. The old snow has larger values of $\langle g_0 \rangle$ (e.g., around 0.8). The pair (p, q)

can be derived from the fitting results derived using geometrical optics approach for particles of various shapes and their mixtures.

It follows for the probability of photon absorption of large nonspherical partilces:

$$\beta = \frac{1}{2}(1 - \rho)(1 - \exp(-\alpha p \overline{D})). \tag{2.99}$$

Snow grains are weakly absorbing in the visible and near infrared. Then it follows:

$$< C_{abs} >= B\alpha\overline{V}, \tag{2.100}$$

where the parameter B depends on the shape of particles.

2.2.2 Integral Light Scattering and Absorption Characteristics of Snowpack

The snow is composed of particles having different shapes, sizes, and orientations. Therefore, the snow scattering, absorption, extinction coefficients and asymmetry parameter are defined as

$$k_{sca} = N\overline{C}_{sca}, k_{abs} = N\overline{C}_{abs}, k_{ext} = N\overline{C}_{ext}, g = \frac{\overline{gC_{sca}}}{\overline{C_{sca}}}. \tag{2.101}$$

Here, N is the number of particles in unit volume of snow and line above the symbol means the averaging with respect to the geometrical parameters and orientation of particles. Equations (2.101) can be used for calculations of other integral light scattering and absorption characteristics useful for studies of radiative transfer in snow layers. They are listed in Table 2.1. The snow optical thickness (SOT) τ is

Table 2.1 Local optical characteristics. The light diffusion parameters listed in the last three lines are derived for weakly absorbing strongly light scattering media ($\omega_0 \rightarrow 1$)

Single scattering albedo	ω_0	$\frac{k_{sca}}{k_{ext}}$
Probability of photon absorption	β	$1 - \omega_0$
Symmetry parameter	ϵ	$1-g$
Transport extinction coefficient	k_{tr}	$(1-g) k_{sca} + k_{abs}$
Diffusion exponent	$k(\omega_0 \rightarrow 1)$	$\sqrt{3(1 - \omega_0)(1 - \omega_0 g)}$
Diffusion extinction coefficient	$\gamma(\omega_0 \rightarrow 1)$	kk_{ext}
Diffusion similarity parameter	$s(\omega_0 \rightarrow 1)$	$k/3(1 - g)$

defined as: $\tau = k_{ext}h$, where h is the snow geometrical thickness and it is assumed that k_{ext} does not change along the vertical coordinate z. Otherwise, it follows:

$$\tau = \int_{z_1}^{z_2} k_{ext}(z)dz, \qquad (2.102)$$

where z_2 gives the position of the upper snow boundary, z_1 is the vertical coordinate of the lower snow boundary and, therefore, $h = z_2 - z_1$.

Introducing the ice grain volume concentration $c = N\overline{V}$, we derive:

$$k_{ext} = \frac{3c}{\overline{D}}, \qquad (2.103)$$

$$k_{abs} = \frac{3c}{2\overline{D}}(1 - \rho)(1 - \exp(-\alpha p\overline{D})) \qquad (2.104)$$

and $k_{sca} = k_{ext} - k_{abs}$.

The volumetric concentration of ice particles in snow is often close to 1/3. Then it follows that $k_{ext} = \frac{1}{\overline{D}}$. Taking into account that the effective diameter of grains is usually in the range 0.1-1 mm and k_{ext} is in the range 1–10 mm^{-1}, we obtain that snow optical thickness is usually larger than 5, where we assumed that the snow layer thickness $h > 5$ mm. The snow with the depth of 10 cm (or even smaller than that for the fine-grained snow) has optical thickness above 100 and, therefore, in a good approximation such a snow layer can be considered as a semi-infinite medium in the visible range. This thickness is even smaller in the near infrared range, where ice also absorbs light. Therefore, photons have smaller probability to survive during propagation in snowpack. Such a behavior of snowpack has important implications as far as radiative transfer in snow is concerned. Firstly, the direct light beam attenuates very fast and can be neglected at depth 5 mm or so. Therefore, the internal light field in snow originates mostly from the diffuse light, which attenuation coefficient γ is much smaller as compared to the extinction coefficient k_{ext}. Secondly, for most of applications snow can be considered as a semi-infinite medium in the optical range. Therefore, the influence of underlying surface on snow reflectance can be neglected in most of cases. Thin layers of snow (5 mm or so) can be considered as turbid layers with large optical thickness. Therefore, simplified asymptotic solutions of the radiative transfer equation valid at large value of τ (see below) can be used.

It follows in the visible and near infrared (till roughly 1.24 μm for natural snow) as discussed above:

$$k_{abs} = B\alpha c. \qquad (2.105)$$

In this range snow is weakly absorbing and main equations simplify. In particular, one can neglect the dependence of the asymmetry parameter $\langle g \rangle$ on the snow grain size and refractive index due to small variation of n and also because $\alpha\overline{D} \to 0$. Then it follows:

$$\langle g \rangle \approx \langle g_0 \rangle. \tag{2.106}$$

Also one derives as $\alpha \overline{D} \to 0$:

$$\beta = \frac{1}{3} \alpha B \overline{D}, \tag{2.107}$$

$$k = \sqrt{\alpha B (1 - g_0) \overline{D}}, \tag{2.108}$$

$$s = \frac{1}{3} \sqrt{\frac{\alpha B \overline{D}}{(1 - g_0)}}, \tag{2.109}$$

where the parameters β, κ, s are defined in Table 2.1. One can also introduce the asymptotic flux attenuation coefficient (AFEC)

$$\gamma = k k_{ext} \tag{2.110}$$

and transport extinction coefficient for nonabsorbing snow (at $k_{abs} < < k_{ext}$):

$$k_{tr} = (1 - g_0) k_{ext}. \tag{2.111}$$

The transport extinction coefficient k_{tr} determines the light transmittance through finite optically thick nonabsorbing snow layers. It coincides with the radiation pressure coefficient for nonabsorbing media. This is due to the fact that $g_0 k_{sca}$ is proportional to the forward carried momentum in the scattered light. It follows for these parameters in the weak absorption approximation:

$$\gamma = 3c \sqrt{\alpha B (1 - g_0) \overline{D}^{-1}}, \quad k_{tr} = \frac{3(1 - g_0)c}{\overline{D}}. \tag{2.112}$$

The parameters mentioned above describe integral light scattering, extinction and absorption characteristics of snow layers. The e-folding depth $z_e = 1/\gamma$ can be presented as

$$z_e = \frac{1}{3c \sqrt{\alpha B (1 - g_0)/\overline{D}}}. \tag{2.113}$$

The dependence of z_e on the wavelength and size of particles at $c = 1/3$, $g_0 = 3/4$, $B = 1.6$ calculated using imaginary part of ice refractive index derived by Picard et al. (2016) is shown in Fig. 2.11. The value of z_e gives the distance at each the

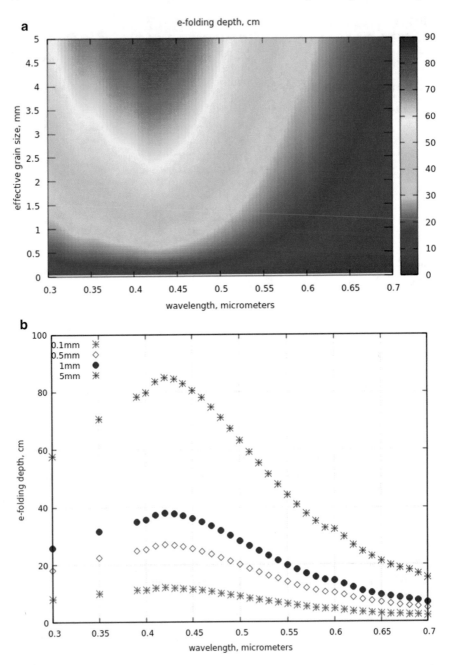

Fig. 2.11 a The dependence of the e-folding depth on the wavelength and effective grain size for clean snow in the visible and near UV. **b** The dependence of the e-folding depth on the wavelength at several values of effective grain size for clean snow in the visible and near UV. **c** The dependence of the e-folding depth on the wavelength at several values of effective grain size for clean snow in the near infrared range of the electromagnetic spectrum

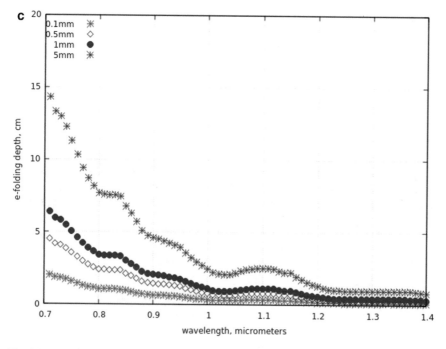

Fig. 2.11 (continued)

diffuse light intensity decreases in e-times (in the deep layer regime). This e-folding depth can be presented as: $z_e(\lambda) = 0.45\sqrt{\lambda \overline{D}/\chi_{ice}(\lambda)}$ for the parameters specified above. Here, $\chi_{ice}(\lambda)$ is the imaginary part of ice refractive index (Picard et al. 2016).

2.2.3 Phase Function

The snow phase function $p(\theta)$ describes angular pattern of light scattered by a unit volume of snow. It can be calculated assuming a particular shape of particles or particular mixtures of different shapes (Liou and Yang 2016). The phase functions of spherical particles are characterized by peaks in the rainbow and glory regions, which absent for snow. Therefore, the assumption of spherical scatterers can not be used for the calculation of snow phase function.

Various parametrizations of snow phase functions have been used in studies of light propagation in snow. In particular, one can use the following parameterization:

$$p(\theta) = \sum_{n=1}^{N} a_n p_n(\theta) \tag{2.114}$$

where

$$p_n(\theta) = \exp(-b_n\theta) \tag{2.115}$$

or

$$p_n(\theta) = \frac{1 - b_n^2}{\left(1 + b_n^2 - 2b_n cos\theta\right)^{\frac{3}{2}}}. \tag{2.116}$$

The function (2.116) is called the Henye-Greenstein (H-G) phase function. Also other parametrizations or combinations of functions given above can be used. In particular, it follows at $N = 1$ from Eq. (2.114), if one uses the function (2.116):

$$p(\theta) = \frac{1 - b_1^2}{\left(1 + b_1^2 - 2b_1 cos\theta\right)^{\frac{3}{2}}}, \tag{2.117}$$

where $b_1 = g$ is the asymmetry parameter. It follows at $N = 2$:

$$p(\theta) = \Upsilon \frac{1 - b_1^2}{\left(1 + b_1^2 - 2b_1 cos\theta\right)^{\frac{3}{2}}} + (1 - \Upsilon) \frac{1 - b_2^2}{\left(1 + b_2^2 - 2b_2 cos\theta\right)^{\frac{3}{2}}}. \tag{2.118}$$

The asymmetry parameter for this phase function is

$$g = \Upsilon b_1 + (1 - \Upsilon)b_2 \tag{2.119}$$

and the backscattering fraction

$$B = \int_{\frac{\pi}{2}}^{\pi} p(\theta) sin\theta d\theta \tag{2.120}$$

is given by the following formula:

$$B = \Upsilon B_1 + (1 - \Upsilon)B_2, \tag{2.121}$$

where

$$B_j = \frac{1 - b_j}{2b_j}\left[\frac{1 + b_j}{\sqrt{1 + b_j^2}} - 1\right]. \tag{2.122}$$

The coefficients Υ, b_1, b_2 in Eq. (2.118) can be selected in such a way that the asymmetry parameter g, the backscattering fraction B and phase function in the backward direction

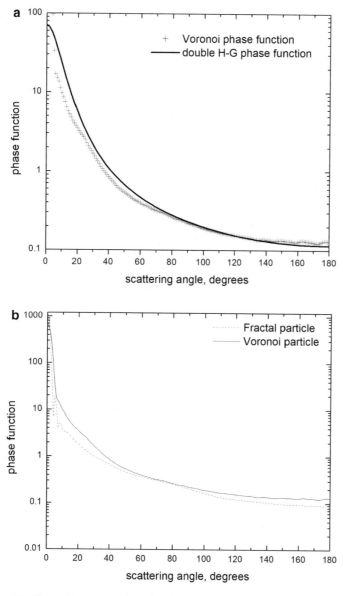

Fig. 2.12 a The Voronoi aggregate phase function (size parameter 311, n = 1.31) and double Henye-Greenstein phase function at $\Upsilon = 0.8615$, $b_1 = 0.85$, $b_2 = 0.2$. **b** Phase function of ice crystals at the wavelength 0.56 μm (n = 1.31) calculated using fractal (tetrahedron of second generation with the side of 100 μm) (Macke et al. 1996) and Voronoi (the size parameter 311) aggregate (Ishimoto et al. 2010) irregularly shaped particles

$$p(\pi) = \Upsilon \frac{1 - b_1}{(1 + b_1)^2} + (1 - \Upsilon) \frac{1 - b_2}{(1 + b_2)^2} \qquad (2.123)$$

coincide with the results derived from the numeric calculations in the framework of ray optics approach for particles of various shapes. In particular, the results derived for the fractal particles (say, Koch fractals of second generation (Macke et al. 1996)), Gaussian random particles (Muinonen 1996), Voronoi (1908) aggregates (Ishimoto et al. 2010, 2018; Letu et al. 2016) and other shapes (see, e.g., Libois et al. 2013, 2014) can be used.

The intercomparison of the double Henyey–Greenstein (1941) phase function and the phase function derived for the Voronoi aggregate is given in Fig. 2.12a. The Voronoi aggregate particles are constructed using the spatial Poisson–Voronoi tessellation to shape modelling of snow grains (Ishimoto et al. 2010, 2018). The discrepancy of Voronoi aggregate and double H-G phase functions is smaller as compared to differences in phase functions for different shapes of particles. In particular, we show the intercomparison of Voronoi and fractal particles in Fig. 2.12b. They differ considerably in the angular range 0–30 degrees. However, the behavior is quite similar in the backward hemisphere relevant to snow remote sensing. One can use the following parameterization of the fractal ice grains phase function, which follows from Eqs. (2.114 and 2.115) at $b_1 = 0.087$ deg^{-1}, $b_2 = 0.014$ deg^{-1}, $a_1 = 11.1$, $a_2 = 1.1$:

$$p(\theta) = 11.1 \exp(-0.087\theta) + 1.1 \exp(-0.014\theta). \qquad (2.124)$$

This equation can be used only outside diffraction peak and do not satisfy the phase function normalization condition.

2.2.4 Polarization Characteristics

Light beam is characterized not just by the intensity, direction of propagation and frequency. It also can be totally or partially polarized. The solar light incident on the top of terrestrial atmosphere is almost unpolarized. However, it can be polarized due to scattering processes in atmosphere and reflection/refraction by natural surfaces including snow. Polarization of incident unpolarized light by snow is rather weak in the visible. Any light beam can be characterized by the Stokes vector-parameter $\vec{S}$ (I, Q, U, V). The parameters I, Q, U, V describe the light intensity (I), degrees of linear (p_e), circular (p_c) and total (p) polarization

$$p_l \equiv \frac{\sqrt{Q^2 + U^2}}{I}, \quad p_c \equiv \frac{V}{I}, \quad p \equiv \sqrt{p_l^2 + p_c^2}. \qquad (2.125)$$

The parameters (Q, U, V) also describe the orientation and ellipticity of polarization ellipse. The transformation of the incident Stokes vector parameter $\vec{S_0}$ (I_0, Q_0, U_0, V_0) by a local snow volume can be presented in the following way:

$$\vec{S} = \Upsilon \widehat{F} \vec{S_0}, \tag{2.126}$$

where $\widehat{F}$ is the scattering matrix, $\Upsilon = (kr)^{-2}, k = \frac{2\pi}{\lambda}$, r is the distance to the observation point. The phase matrix $\widehat{P}$ is proportional to the scattering matrix with the element P_{11} equal to the phase function:

$$\widehat{P} = \frac{4\pi \widehat{F}}{k^2 \overline{C}_{sca}}, \tag{2.127}$$

where $\overline{C}_{sca}$ is the average scattering cross section.

The normalized phase matrix elements are often used. They are defined as follows:

$$p_{ij} = P_{ij}/P_{11}. \tag{2.128}$$

In the case of randomly distributed nonspherical particles the phase matrix has the following general block–diagonal form:

$$\widehat{P} = \begin{pmatrix} P_{11} & P_{12} & 0 & 0 \\ P_{12} & P_{22} & 0 & 0 \\ 0 & 0 & P_{33} & P_{34} \\ 0 & 0 & -P_{34} & P_{44} \end{pmatrix}. \tag{2.129}$$

Therefore, Stokes parameters of scattered light can be presented in the following form:

$$\begin{aligned} I &= P_{11}I_0 + P_{12}Q_0, \\ Q &= P_{12}I_0 + P_{22}Q_0, \\ U &= P_{33}U_0 + P_{34}V_0, \\ V &= -P_{34}U_0 + P_{44}V_0. \end{aligned} \tag{2.130}$$

It follows in the case of incident unpolarized light ($I_0 = 1, Q_0 = U_0 = V_0 = 0$):

$$I = P_{11}, Q = P_{21}, U = 0, V = 0. \tag{2.131}$$

This means that the scattered light becomes linearly polarized with the degree of polarization equal to the value of the normalized phase matrix element $p_{12} = P_{12}/P_{11}$. Also one can easily establish physical meaning of other normalized phase matrix elements. Let us assume that the incident light is left handed circularly polarized ($I_0 = V_0 = 1, U_0 = Q_0 = 0$). Then it follows from Eq. (2.130):

$$I = P_{11}, Q = P_{12}, U = P_{34}, V = P_{44}. \tag{2.132}$$

The degree of circular polarization (DCP) is equal to $p_{44} = P_{44}/P_{11}$. Therefore, the element p_{44} is equal to the degree of circular polarization of singly scattered light under left handed circularly polarized light illumination conditions. Let us consider the illumination of a particle by the horizontally polarized light ($I_0 = -U_0 = 1$, $V_0 = Q_0 = 0$). Then it follows from Eq. (2.130):

$$I = P_{11}I_0, \quad Q = P_{12}, \quad U = 0, \quad V = P_{34}. \tag{2.133}$$

Therefore, the element p_{34} is equal to the degree of circular polarization of singly scattered light under horizontally polarized light illumination conditions. It shows the effectivity of linear to circular light polarization state transformation during single scattering events. In a similar way one can establish the physical meaning of other phase matrix elements.

The normalized phase matrix elements of Voronoi aggregate and fractal particle models derived in the framework of geometrical optics approach are given in Fig. 2.13. The calculations have been performed at the wavelength 0.56 μm, $n = 1.31$ and the size parameter x = 311 (for Voronoi aggregates) and the characteristic length 100 μm (the side of initial tetrahedron) of a Koch fractal of second generation. It follows that the normalized phase matrix elements almost coincide for these diverse two models of ice grain shapes. One can see that such particles only weakly polarize

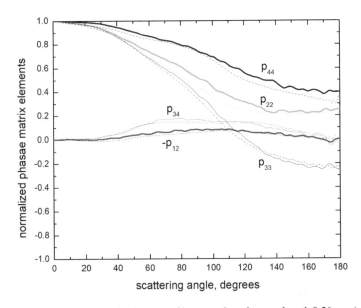

Fig. 2.13 Normalized phase matrix elements of ice crystals at the wavelength 0.56 μm (n = 1.31) calculated using fractal (tetrahedron of second generation with the side of 100 μm) and Voronoi (the size parameter 311) aggregate irregularly shaped particles

incident unpolarized light (the degree of linear polarization p_{12} is smaller than 10% with maximum polarization in the side scattering direction). The value of DCP p_{44} decreases with the scattering angle. The value of the phase matrix element p_{34} (linear to circular polarized light conversion probability) is small for all scattering angles. Such behavior is consistent with measured polarization characteristics of snow layers.

2.2.5 Light Absorption by Polluted Snowpack

Snow can contain various inclusions (soot, dust, algae, litter from the trees, etc.). Pollutants modify snow optical properties and appearance. It is often the case that the concentration of pollutants is too small to influence light scattering processes in snow. However, snow hardly absorbs light in the visible. Therefore, even small concentrations of pollutants influence local absorption coefficient of snow. Assuming external mixture of pollutants and ice grains, one may write for the snow absorption coefficient:

$$k_{abs} = k_{abs}^{ice} + k_{abs}^{pol}, \tag{2.134}$$

where the ice absorption coefficient k_{abs}^{ice} is given by Eq. (2.104) and the absorption coefficient of pollutants k_{abs}^{pol} can be presented as

$$k_{abs}^{pol} = c_p \gamma_p, \tag{2.135}$$

where c_p is the volumetric concentration of pollutants and

$$\gamma_p = \frac{\left\langle C_{abs}^{pol} \right\rangle}{\left\langle V_{pol} \right\rangle}. \tag{2.136}$$

is the volumetric absorption coefficient (VAC) of pollutants. Here, $\left\langle V_{pol} \right\rangle$ is the average volume of impurity particles and $\left\langle C_{abs}^{pol} \right\rangle$ is their average cross section. For very small and also for very large (as compared to the wavelength) weakly absorbing particles, the cross section $\left\langle C_{abs}^{pol} \right\rangle$ is proportional to the product of the bulk absorption coefficients of pollutants $\alpha_{pol} = \frac{4\pi \chi_{pol}}{\lambda}$ and $\left\langle V_{pol} \right\rangle$. Here, χ_{pol} in the imaginary part of impurity complex refractive index $m_{pol} = n_{pol} - i \chi_{pol}$. Then it follows:

$$\gamma_p(\lambda) = B_{pol} \alpha_{pol}(\lambda). \tag{2.137}$$

In particular, it follows at the size of pollutants much smaller than the wavelength:

$$B_{pol} = \frac{9n_{pol}}{\left(2 + n_{pol}^2\right)\left(2 + n_{pol}^2 + \chi_{pol}^2\right) + \chi_{pol}^4}. \tag{2.138}$$

Assuming that the soot complex refractive index is 1.95–0.79i across solar spectrum, one derives: $B_{pol} = 1.3$.

One can derive the following simple equation for snow contaminated by soot:

$$k_{abs} = c_{ice}B_{ice}\alpha_{ice} + c_{pol}B_{pol}\alpha_{pol} \tag{2.139}$$

valid in the visible and near infrared (at small values of the probability of photon absorption).

In general, the value of $\left\langle C_{abs}^{pol}\right\rangle$ can be derived averaging the absorption cross section of pollutants C_{abs}^{pol} with respect to their size and shape distribution. It follows for the spherical particles:

$$< C_{abs}^{pol} \geq \int_0^\infty C_{abs}^{pol}(a)f(a)da, \tag{2.140}$$

where $f(a)$ is the size distribution of impurities. One can see that the calculation of $\left\langle C_{abs}^{pol}\right\rangle$ requires information not only on the complex refractive index of impurity particles but also their size distribution must be known. This is often not the case. Therefore, various approximations of the VAC γ_p is used. In particular, one can assume that

$$\gamma_p(\lambda) = \gamma_p(\lambda_0)\left(\frac{\lambda}{\lambda_0}\right)^{-p}. \tag{2.141}$$

One can also define the mass absorption coefficient (MAC) of pollutants:

$$M_p(\lambda) = \frac{\gamma_p}{\rho}. \tag{2.142}$$

Here ρ is the density of an impurity particle (1.8 g/cm^3 for black carbon (BC) or soot, 2.5 g/cm^3 for dust). The black carbon MAC $M_p(\lambda)$ decreases from 6.8 $\frac{m^2}{g}$ at 0.4 μm to 3.6 $\frac{m^2}{g}$ at 1 μm. It is equal to 0.7 $\frac{m^2}{g}$ at 4 μm.

The assumption of the external mixture of pollutants gives the lower boundary of the influence of pollutants on the snow absorption. As a matter of fact, some pollutants are contained inside snow grains (internal mixture). This leads to even greater influence of pollutants on snow absorption due to focusing properties of snow grains. The influence of pollutants on snow absorption is at maximum in the visible, where the maximum of the incident solar flux is located. It generally diminishes with the wavelength and can be neglected at channels, where ice strongly absorbs solar light (Dombrovsky and Kokhanovsky 2021).

References

Dombrovsky, L., and A.A. Kokhanovsky. 2021. Solar heating of cryosphere: Snow and ice sheets. *Springer Series in Light Scattering* 7: 53–109.

Fresnel A. 1831. *Mémoire sur la loi des modifications que la réflexion imprime à la lumière polarisée* (read 7 January 1823).

Garrett, T.J., P.V. Hobbs, and H. Gerber. 2001. Shortwave, single-scattering properties of arctic ice clouds. *Journal of Geophysical Research* 106: 15155–15172.

Henyey, L.G., and J.L. Greenstein. 1941. Diffuse radiation in the Galaxy. *Astrophysical Journal* 93: 70–83.

Ishimoto, H., et al. 2010. Shape modeling of mineral dust particles for light scattering calculations using the spatial Poisson–Voronoi tessellation. *JQSRT* 111: 2434–2443.

Ishimoto, H., et al. 2018. Snow particles extracted from X-ray computed microtomography imagery and their single scattering properties. *JQSRT* 209: 113–128.

Kokhanovsky, A.A., and E.P. Zege. 1995. Local optical parameters of spherical polydispersions: Simple approximations. *Applied Optics* 34: 5513–5519.

Libois, Q., et al. 2013. Influence of grain shape on light penetration in snow. *The Cryosphere* 7: 1803–1818.

Libois, Q., M. Dumont, G. Picard, and L. Arnaud. 2014. Experimental determination of the absorption enhancement parameter of snow. *Journal of Glaciology* 60: 714–724.

Liou, K.-N., and P. Yang. 2016. *Light scattering by ice crystals: Fundamentals and applications.* Cambridge: Cambridge University Press.

Letu, H., et al. 2016. Ice cloud properties from Himawari-8/AHI next generation geostationary satellite: Capability to the AHI to monitor the cloud generation process. *IEEE Transactions on Geoscience and Remote Sensing* 57: 3219–3239.

Macke, A., J. Mueller, and E. Raschke. 1996. Single scattering properties of atmospheric ice crystals. *Journal of Atmospheric Science* 52: 2813–2825.

Maxwell, J.C. 1873. *A treatise on electricity and magnetism.* Oxford.

Mie, G. 1908. Beiträge zur Optik trüber Medien, speziell kolloidaler Metallösungen. *Annalen Der Physik, Vierte Folge* 25 (3): 377–445.

Muinonen, K. 1996. Light scattering by Gaussian random particles: Ray optics approximation. *JQSRT* 55: 577–601.

Picard, G., Q. Libois, and L. Arnaud. 2016. Refinement of the ice absorption spectrum in the visible using radiance profile measurenst in Antarctic snow. *The Cryosphere* 10: 2655–2672.

Shifrin, K.S. 1951. *Scattering of light in a turbid medium.* Leningrad: Gostekhteorizdat.

van de Hulst, H.C. 1957. *Light scattering by small particles.* London: Chapman and Hill.

Voronoi, G.F. 1908. Nouvelles applications des paramètres continus à la théorie de formes quadratiques. *Journal Für Die Reine Und Angewandte Mathematik* 134: 198–287.

Zhou, X., S. Li, and K. Stamnes. 2003. Geometrical–optics code for computing the optical properties of large dielectric spheres. *Applied Optics* 42: 4295–4306.

Chapter 3
Radiative Transfer in Snowpack

3.1 Radiative Transfer Characteristics

The main task of this chapter is to derive simple analytical equations, which govern light intensity and fluxes inside snowpack and also at its boundaries. The spectral light intensity I_λ is defined as the amount of radiant energy dW in the wavelength interval $[\lambda, \lambda + d\lambda]$ transported across an element of area dS into a solid angle $d\Omega$ oriented with an angle ϑ relative to the normal to the area dS, within a time interval dt:

$$I_\lambda = \frac{dW}{\cos \vartheta \, d\Omega \, d\lambda \, dS \, dt}.$$

(3.1)

Therefore, if one derives the spatial and angular distribution of the light intensity I_λ, the amount of radiation energy dW transported through an element of area dS into a solid angle $d\Omega$ in the spectral interval $d\lambda$ and time interval dt in snow can be calculated as

$$dW = I_\lambda \cos \vartheta \, d\Omega \, d\lambda \, dS \, dt.$$

(3.2)

It is useful to introduce irradiance as the amount of energy transported through an element of area dS in in the wavelength interval $d\lambda$ and time interval dt

$$dE_\lambda = \frac{dW}{d\lambda \, dS \, dt}.$$

(3.3)

The radiant power

$$d\Phi = \frac{dW}{dt}$$

(3.4)

© Springer Nature Switzerland AG 2021
A. Kokhanovsky, *Snow Optics*,
https://doi.org/10.1007/978-3-030-86589-4_3

is the amount of transported energy in unit time. It is measured in watts (J/s). The irradiance is measured in $\left[\text{W} \cdot \text{m}^{-2} \cdot \text{nm}^{-1}\right]$.

An important point is that the monochromatic irradiance dE_λ can be calculated as [see Eqs. (3.2), (3.3)]:

$$dE_\lambda(\vec{r}) = I_\lambda\left(\vec{r}, \vec{\Omega}\right) \cos \vartheta \, d\vec{\Omega}, \tag{3.5}$$

where the light intensity $I_\lambda\left(\vec{r}, \vec{\Omega}\right)$ is measured in $\left[\text{W} \cdot \text{m}^{-2} \cdot \text{sr}^{-1} \cdot \text{nm}^{-1}\right]$. It is a function of coordinates $\vec{r}$, propagation direction $\vec{\Omega}$ and time t:

$$I_\lambda = I_\lambda(\vec{r}, \Omega, t), \tag{3.6}$$

where $\vec{r}$ is the position vector incorporating coordinates, $\vec{\Omega}$ is the unit vector in the direction of light propagation, and t is the time. The intensity does not change with time assuming that the properties of the light propagation channel (say, a turbid medium) and light source do not change in time (stationary media and light sources). One should account for the time dependence of light intensity, e.g., in the case of problems related to the illumination of snow by a pulsed (say, laser) light source. In this book we shall consider only stationary light fields and turbid media.

For the medium stratified in plane-parallel horizontally homogeneous layers such as a snowpack, the intensity depends on the vertical coordinate z and the propagation direction $\vec{\Omega}$:

$$I_\lambda = I_\lambda\left(z, \vec{\Omega}\right) = I_\lambda(z, \mu, \varphi). \tag{3.7}$$

Here the direction $\vec{\Omega} = \{\mu, \varphi\}$ is defined in spherical coordinates through the cosine μ of the zenith angle θ and azimuthal angle φ, $\mu > 0$ for downwelling radiation and $\mu < 0$ for the upwelling radiation. Note that

$$d\Omega = d\mu d\varphi. \tag{3.8}$$

If the light intensity is independent on direction, then it is called isotropic, that is

$$I_\lambda = I_\lambda(z). \tag{3.9}$$

If the intensity is integrated over all wavelengths, it is called the integrated intensity and given by

$$I = \int_0^\infty I_\lambda d\lambda. \tag{3.10}$$

The irradiances for the downward and upward propagating light are defined as

$$F_\lambda^\downarrow(z) = \int_0^{2\pi} \int_0^1 I_\lambda(z, \mu, \varphi) \mu \, d\mu \, d\varphi \qquad (3.11)$$

and

$$F_\lambda^\uparrow(z) = \int_0^{2\pi} \int_{-1}^0 I_\lambda(z, \mu, \varphi) \mu \, d\mu \, d\varphi. \qquad (3.12)$$

Quantities (3.11), (3.12) are called downwelling and upwelling light fluxes, respectively. The integrated irradiance over all wavelengths reads as

$$F = \int_0^\infty F_\lambda d\lambda. \qquad (3.13)$$

Yet another quantity of interest is the radiation energy density. It is defined as

$$\rho_\lambda = \frac{1}{c} \int_{4\pi} I_\lambda d\Omega, \qquad (3.14)$$

where c is the light speed. The actinic flux

$$\Phi_a = c\rho_\lambda \qquad (3.15)$$

is needed for the calculation of absorbed radiation power at a given location in snow. It particular, this radiative characteristic can be used for the studies of photodissociation rates of various molecules in snow.

One can see that the light intensity I_λ is the main radiative transfer characteristic, which defines all other optical properties (e.g., radiant power, irradiances, actinic fluxes, radiation energy density, etc.). In the next section we shall introduce the equation, which can be used to calculate the light intensity for plane parallel horizontally homogeneous snowpacks uniformly illuminated at a top by a plane-parallel light beam.

It should be pointed out that the intensity I_λ is related to the photon distribution function $f_\lambda(\mathbf{r}, \boldsymbol{\Omega})$ by a simple equation:

$$I_\lambda = ch\nu f_\lambda(\mathbf{r}, \mathbf{\Omega}), \tag{3.16}$$

where $\epsilon = h\nu$ is the energy of a photon, h is the Planck constant and $\nu = c/\lambda$ is the frequency. The integral $N(\vec{r}) = \int_{4\pi} f_\lambda(\vec{r}, \overrightarrow{\Omega})d\overrightarrow{\Omega}$ gives the number of photons at the position specified by the vector $\vec{r}$. Equation (3.16) follows from the comparison of Eq. (3.2) with yet another representation of the radiant energy:

$$dW = \epsilon dN, \tag{3.17}$$

where

$$dN = f_\lambda\left(\vec{r}, \overrightarrow{\Omega}\right)\cos\vartheta d\Omega d\lambda dV \tag{3.18}$$

is the number of photons with wavelengths in the interval $[\lambda, \lambda + d\lambda]$ located in time t in a volume $dV = cdtdS$ near point $\vec{r}$ and moving within a solid angle $d\Omega$ around direction $\overrightarrow{\Omega}$. It follows from Eqs. (3.14), (3.16) for the radiation energy density:

$$\rho_\lambda = h\nu\int_{4\pi} f_\lambda\left(\vec{r}, \overrightarrow{\Omega}\right)d\Omega. \tag{3.19}$$

Therefore, the ratio

$$N = \frac{\rho_\lambda}{h\nu} \tag{3.20}$$

gives the number of photons with the frequency ν in unit volume of a snowpack. This parameter is of importance for the snow photochemistry.

3.2   Radiative Transfer Equation

The snow radiative characteristics can be derived using the radiative transfer equation (RTE) for the diffuse light intensity I. This equation can be written in the following form for the plane-parallel snow layers (Chandrasekhar 1950):

$$\cos\vartheta\frac{dI(\vartheta, \varphi)}{d\tau} = -I(\vartheta, \varphi) + \frac{\omega_0}{4\pi}\int_0^{2\pi} d\varphi'\int_0^\pi d\vartheta' p(\vartheta', \varphi' \to \vartheta, \varphi)I(\vartheta', \varphi')$$

$$+ \frac{\omega_0}{4} p(\vartheta_0, \varphi_0 \to \vartheta, \varphi)F_0 e^{-\frac{\tau}{\cos\vartheta_0}}. \tag{3.21}$$

Here ω_0 is the single scattering albedo, τ is the optical thickness, $p(\vartheta', \varphi' \to \vartheta, \varphi)$ is the phase function, πF_0 is the incident solar flux at the unit area perpendicular to the light beam, ϑ_0 is the solar incidence angle, φ_0 is the solar azimuthal angle, ϑ is the viewing zenith angle, and φ is the viewing azimuthal angle. This integro—differential equation enables the calculation of the diffuse light field intensity at any level and direction of propagation inside of the snow layer and also at the upper and lower boundaries of a snowpack. The calculation must be done taking into account boundary conditions. The simplest approximation is to assume that there is no diffuse light incident on the snow layer from above and also from below (underlying black surface). Such a condition is close to reality, e.g., in Antarctica, where atmospheric optical thickness is very small and diffuse light coming from the sky can be often neglected (say, at near infrared region of electromagnetic spectrum). For polluted atmosphere, the account for the diffuse light incident from the sky is of importance. In this case one needs to specify the vertical distribution of local optical properties of atmosphere—snow system (Aoki et al. 1999) and solve Eq. (3.1) numerically (say, in the framework of the discrete ordinates technique).

The direct light beam intensity unlike diffuse light intensity does not require complex numerical calculations. Namely, it follows:

$$I_{dir} = F_0 \exp\left(-\frac{\tau}{\cos \vartheta_0}\right) \delta(\vec{\Omega} - \vec{\Omega}_0), \qquad (3.22)$$

where the delta function $\delta(\vec{\Omega} - \vec{\Omega}_0)$ signifies the propagation direction $\left(\vec{\Omega} = \vec{\Omega}_0\right)$ of the direct beam. It follows that the direct beam attenuates faster for larger solar zenith angles ϑ_0.

The integral in Eq. (3.21) accounts for multiple scattering effects. The single scattering contribution to the reflected and transmitted light can be derived from Eq. (3.21) ignoring the integral term. The remaining differential equation can be solved analytically. The answer is:

$$I(x) = \frac{F_0}{4} e^{-x} \int_{\psi}^{x} \omega_0(x') p(x') \exp(-(q-1)x'), \qquad (3.23)$$

where $x = \tau/\cos \vartheta$, $q = \cos \vartheta/\cos \vartheta_0$. The constant ψ can be found from the boundary conditions. Let us assume that the diffuse light intensity incident both on the top and bottom of a scattering layer is equal to zero:

$$I_\downarrow(x = 0) = 0, \ \cos\vartheta > 0, \ \ I_\uparrow(x = x_0) = 0, \ \cos\vartheta < 0. \qquad (3.24)$$

Here arrows show the direction of light propagation, $x = 0$ corresponds to the upper boundary, $x = x_0$ corresponds to the lower boundary of the scattering layer with the optical thickness τ_0 and $x_0 = \tau_0/\cos \vartheta$. Let us assume that the local optical properties do not change with the depth. Then the integral (3.23) can be evaluated analytically:

$$I_{\downarrow}(x) = \frac{\omega_0 p(\theta) F_0}{4(q-1)} \left\{ e^{-x} - e^{-qx} \right\}, \tag{3.25}$$

$$I_{\uparrow}(x) = \frac{\omega_0 p(\theta) F_0}{4(q-1)} \left\{ e^{-x-(q-1)x_0} - e^{-qx} \right\}, \tag{3.26}$$

where θ is the scattering angle and we have accounted for the boundary conditions. It follows at the boundaries:

$$I_{\downarrow}(x_0) = \frac{\omega_0 p(\theta) F_0}{4(q-1)} \left\{ e^{-x_0} - e^{-qx_0} \right\}, \tag{3.27}$$

$$I_{\uparrow}(0) = \frac{\omega_0 p(\theta) F_0}{4(q-1)} \left\{ e^{-(q-1)x_0} - 1 \right\}. \tag{3.28}$$

Let us introduce reflection and transmission functions:

$$R(\mu, \varphi, \mu_0, \varphi_0) = \frac{\pi I_{\uparrow}(\vartheta, \varphi, \tau = 0)}{F_0 \xi}, \tag{3.29}$$

$$T(\mu, \varphi, \mu_0, \varphi_0) = \frac{\pi I_{\downarrow}(\vartheta, \varphi, \tau = \tau_0)}{F_0 \xi}, \tag{3.30}$$

where $\xi = \cos \vartheta_0$. Then it follows:

$$R = \frac{\omega_0 p(\theta)}{4(\xi + \eta)} \left\{ 1 - \exp\left[\left(\frac{1}{\xi} + \frac{1}{\eta} \right) \tau_0 \right] \right\}, \tag{3.31}$$

$$T = \frac{\omega_0 p(\theta)}{4(\xi - \eta)} \left\{ \exp\left[-\frac{\tau_0}{\eta} \right] - \exp\left[-\frac{\tau_0}{\xi} \right] \right\}, \tag{3.32}$$

where $\eta = |\cos \vartheta|$ and $\cos \theta = -\xi\eta + \sqrt{(1 - \xi^2)(1 - \eta^2)} \cos(\varphi - \varphi_0)$.
Equation (3.32) has the following form at $\xi = \eta$:

$$T = \frac{\omega_0 \tau_0 p(\theta)}{4\xi^2} \exp\left[-\frac{\tau_0}{\xi} \right]. \tag{3.33}$$

The contribution of single scattering events to the transmission can be neglected for very large values of τ_0 typical for snow layers. However, the contribution of singly scattered photons to the reflection function can not be neglected even for the case of semi-infinite layers and this contribution is given by the following equation [see

Eq. (3.31)]:

$$R_{ss} = \frac{\omega_0 p(\theta)}{4(\xi + \eta)}.$$ (3.34)

This equation shows that not only integral scattering and absorption characteristics but also snow phase function must be accurately modelled to predict snow reflective properties (e.g., angular dependence of snow reflection function). The contribution of multiply scattered photons to the reflection function is discussed in Sect. 3.4.

3.3 Light Field in Deep Layers of Semi-Infinite Weakly Absorbing Snowpack

Let us assume that the medium under consideration is a semi-infinite one and the light intensity at a large optical depth inside the medium is to be derived. In this case one may neglect the contribution of the direct light beam. Then it follows:

$$\mu \frac{dI(\mu)}{d\tau} = -I(\mu) + \frac{\omega_0}{2} \int_{-1}^{1} I(\mu', \mu) d\mu' \overline{p}(\cos\theta),$$ (3.35)

where $I(\mu)$ is the diffused light intensity in the direction specified by the angle $\vartheta = \text{acos}(\mu)$ from the normal to the layer and

$$\overline{p}(\cos\theta) = \frac{1}{2\pi} \int_{0}^{2\pi} p(\cos\theta) d\varphi$$ (3.36)

is the azimuthally averaged phase function.

The intensity in deep snow layers does not change its angular pattern and is attenuated according to the exponential law (Sobolev 1975):

$$I = Ci(\mu) \exp(-k\tau),$$ (3.37)

where k is the diffusion exponent, C is the constant.

Let us substitute this solution to Eq. (3.35). Then it follows under assumption of isotropic scattering ($p = 1$):

$$i(\mu) = \frac{1}{1 - k\mu},$$ (3.38)

where we used the normalization condition:

$$\frac{\omega_0}{2} \int_{-1}^{1} i(\mu')d\mu' = 1. \qquad (3.39)$$

Substituting Eq. (3.38) to equation Eq. (3.39) given above makes it possible to derive the following equation for the diffusion exponent:

$$\omega_0 s(k) = 1, \qquad (3.40)$$

where

$$s(k) = \frac{1}{2k} \ln \frac{1+k}{1-k}. \qquad (3.41)$$

Let us take into account that it follows for weakly absorbing media ($k \to 0$):

$$\ln(1+k) = k - \frac{k^2}{2} + \frac{k^3}{3}. \qquad (3.42)$$

Therefore, one derives:

$$s(k) = 1 + \frac{k^2}{3} \qquad (3.43)$$

and [see Eq. (3.20)]

$$1 + \frac{k^2}{3} = \frac{1}{1-\beta}, \qquad (3.44)$$

where $\beta = 1 - \omega_0$ is the probability of photon absorption. Finally, we derive for weakly absorbing media ($\beta \to 0$) from Eq. (3.40) (at $p = 1$):

$$k = \sqrt{3\beta}. \qquad (3.45)$$

This equation shows how the diffusion exponent depends on PPA for the case of weakly absorbing isotropically scattering media. Therefore, one derives at $p = 1$ and $k \to 0$:

$$i(\mu) = 1 + \sqrt{3\beta}\mu. \qquad (3.46)$$

The phase function of snow is not isotropic and asymmetry parameter is close to 0.75 in the visible. Therefore, it is of importance to consider more complex phase functions. This can be done solving Eq. (3.35) numerically for the phase functions specified in the previous Chapter. However, our task is to derive approximate analytical equations. Therefore, we consider now the case of the following phase function:

$$p(\theta) = 1 + x_1 \cos \theta, \tag{3.47}$$

where $x_1 = 3g$, g is the average cosine of the scattering angle. Then one derives assuming that the value of I is given by Eq. (3.37):

$$(1 - k\mu)i(\mu) = \frac{\omega_0}{2} \int_{-1}^{1} (1 + x_1\mu\mu')i(\mu')d\mu', \tag{3.48}$$

where we accounted for the fact that the azimuthally averaged cosine of scattering angle can be presented as $\langle \cos \theta \rangle_\varphi = \mu\mu'$. Therefore, we derive:

$$i(\mu) = \frac{b(\mu)}{1 - k\mu}, \tag{3.49}$$

where

$$b(\mu) = \frac{\omega_0}{2} \int_{-1}^{1} (1 + x_1\mu\mu')i(\mu')d\mu' \tag{3.50}$$

or

$$b(\mu) = \frac{\omega_0}{2} \int_{-1}^{1} \frac{(1 + x_1\mu\mu')b(\mu')}{1 - k\mu'}d\mu'. \tag{3.51}$$

It is clear from Eq. (3.50) that $b(\mu)$ can be presented in the following form:

$$b(\mu) = b_0 + b_1\mu, \tag{3.52}$$

where

$$b_0 = \frac{\omega_0}{2} \int_{-1}^{1} \frac{b(\mu')d\mu'}{1 - k\mu'}, b_1 = \frac{\omega_0 x_1}{2} \int_{-1}^{1} \frac{\mu'b(\mu')}{1 - k\mu'}d\mu'. \tag{3.53}$$

It follows from Eqs. (3.52), (3.53):

$$b_0 + b_1\mu = b_0 j_0 + b_1 j_1 + b_0 x_1 j_1 \mu + b_1 x_1 j_2 \mu \tag{3.54}$$

or

$$b_0 = b_0 j_0 + b_1 j_1, \quad b_1 = b_0 x_1 j_1 + b_1 x_1 j_2, \tag{3.55}$$

where

$$j_n = \frac{\omega_0}{2} \int_{-1}^{1} \frac{(\mu')^n d\mu'}{1 - k\mu'} \tag{3.56}$$

or after integration

$$j_0 = \omega_0 s, \ j_1 = \frac{\omega_0(s-1)}{k}, \ j_2 = \frac{j_1}{k}. \tag{3.57}$$

Therefore, we have the following equation for the determination of both b_0 and b_1:

$$\hat{a}\vec{b} = 0, \tag{3.58}$$

where the elements of the matrix $\hat{a}$ can be represented in the following way [see Eqs. (3.35), (3.37)]:

$$a_{11} = \frac{1}{\omega_0} - s, \quad a_{12} = a_{21} = \frac{1-s}{k}, \quad a_{22} = \frac{1}{\omega_0 x_1} - \frac{s-1}{k^2}. \tag{3.59}$$

This equation has a solution, if the determinant

$$D = a_{11}a_{22} - a_{12}a_{21} \tag{3.60}$$

is equal to zero. Therefore, we have:

$$\omega_0 \left[1 + x_1 \frac{1-\omega_0}{k^2} \right] s - \omega_0 x_1 \frac{1-\omega_0}{k^2} = 1. \tag{3.61}$$

Equation (3.61) can be used to determine the diffusion exponent at any values of the pair (x_1, ω_0). Let us assume that the medium under consideration is weakly absorbing. Then the probability of photon absorption $\beta = 1 - \omega_0 \rightarrow 0, s \rightarrow 1 + \frac{k^2}{3}$ and, therefore, one derives:

$$k = \sqrt{3\beta(1-g)}, \tag{3.62}$$

which is reduced to the value derived for the case of isotropic scattering in the case $g = 0$ as it should be. Importantly, although this expression for the diffusion exponent in weakly absorbing medium has been derived for the case of a specific phase function, it can be used for any phase functions under assumption that $\beta \to 0$, which is valid in particular for snow in the visible and near infrared. Assuming that $g = 3/4$ for snow in the visible and near infrared, we derive: $k = \sqrt{3\beta}/2$ for snow in the visible and near-infrared. One can see that the diffusion exponent is two times smaller for snow as compared to the case of isotropic scattering under approximation studied. Therefore, the radiation penetration in snow is larger as compared to the case of isotropic scattering for a given value of probability of photon absorption.

It follows for the function $i(\mu)$

$$i(\mu) = b_0 \frac{1 + \xi\mu}{1 - k\mu}, \tag{3.63}$$

where $\xi = b_1/b_0$. This function can be multiplied by any constant and still be a solution of Eq. (3.35). Let us assume that $b_0 = 1$ and $\xi = b_1$. Then it follows from Eq. (3.55):

$$\xi = x_1 j_1 \left(1 + \frac{\xi}{k} \right) \tag{3.64}$$

or

$$\xi = \frac{\Pi}{1 - \Pi/k}, \tag{3.65}$$

where $\Pi = (s - 1)\omega_0 x_1 k^{-1}$. Let us assume that $\beta \to 0$. Then one derives:

$$\xi = \frac{gk\omega_0}{1 - g\omega_0} \tag{3.66}$$

and

$$i(\mu) = 1 + 3s\mu, \tag{3.67}$$

where

$$s = \frac{\beta}{k} \tag{3.68}$$

is the similarity parameter. One can see that the ratio of the probability of photon absorption to the diffusion exponent is a key parameter of the radiative transfer in snow. Also we can write:

$$s = \sqrt{\frac{\beta}{3(1-g)}} \equiv \frac{k}{3(1-g)}. \tag{3.69}$$

The parameters (k, s) can be used to derive the intensity of light in deep layers of weakly absorbing media with the phase function given by Eq. (3.24). Namely, it follows:

$$I(\mu) = (1 + 3s\mu)\exp(-k\tau) \tag{3.70}$$

as $\beta \to 0$. This equation can be also written in the following form:

$$I(\mu) = (1 + 3s\mu)\exp(-\gamma h), \tag{3.71}$$

where

$$\gamma = kk_{ext} \tag{3.72}$$

is the so-called asymptotic flux attenuation coefficient (AFEC) and h is the geometrical depth. It follows from Eq. (3.71) that the ratio ζ of light intensities/fluxes at two depths (h_1, h_2) in deep layers of snow can be presented as

$$\zeta = \exp(-\gamma\Delta), \tag{3.73}$$

where $\Delta = h_2 - h_1$. Therefore, the AFEC can be easily measured experimentally:

$$\gamma = \frac{\ln(F(h_1)/(F(h_2))}{\Delta}. \tag{3.74}$$

Here, $F(h_j)$ is the flux at the depth h_j. Equation (3.74) is also valid for light intensities in a given direction as well. The measurements of light intensities at two directions at a given depth in snow can be used to estimate the parameter s [see Eq. (3.71)]. Although Eqs. (3.62), (3.70) have been derived for a particular phase function, they play a fundamental role in the radiative transfer in weakly absorbing media with arbitrary phase functions. As a matter of fact, they are valid approximations for the snow in the visible and near infrared. Our derivations show that the most important single local optical characteristic of snow is the similarity parameter s. This parameter can be used to find the diffusion exponent $k = 3(1 - g)s$ (under assumption that g is known). Also light intensity distribution in deep snow layers [see Eq. (3.70)] can be derived. The parameter s depends on the ratio of absorption $l_{abs} = 1/k_{abs}$ and transport extinction $l_{tr} = 1/(1 - g)k_{ext}$ (valid at $k_{abs} \ll k_{sca}$) lengths:

$$s = \sqrt{\frac{l_{tr}}{3l_{abs}}}. \tag{3.75}$$

Also it follows: $l_{tr}/l_{abs} = 3s^2$. Generally speaking, the two local optical parameters (γ, s) are needed to describe the light propagation in snow layers. The value of *the diffusion exponent* can be derived from the AFEC in case the snow extinction coefficient k_{ext} is known [see Eq. (3.72)]. We shall see later that the similarity parameter s can be derived not only from the analysis of the angular distribution of transmitted light in deep snow layers but also it can be obtained from the measurements of snow diffuse reflection coefficient under diffuse illumination conditions. This means that the transport extinction length can be also derived:

$$l_{tr} = \frac{3s}{\gamma}. \tag{3.76}$$

In addition, the absorption length is given by the following equation:

$$l_{abs} = \frac{l_{tr}}{3s^2}. \tag{3.77}$$

This opens ways to the reflectance spectroscopy of snow (in particular, the determination of pollution type and load from measurements of snow reflectance spectra).

3.4 Reflection of Light from a Semi-Infinite Snow Layer

3.4.1 Nonlinear Integral Equation for the Reflection Function

Equation (3.1) can be used for any optical thickness. In many cases snow can be considered as a semi-infinite medium. Then it is useful to use the following representation for the reflection function:

$$R = R_{ss} + R_{ms}. \tag{3.78}$$

The value of R_{ss} gives the contribution of single scattering [see Eq. (3.31)] and R_{ms} describes the contribution of multiple scattering to the reflected light intensity.

The value of R_{ms} can be calculated using Eq. (3.1) or alternatively using this nonlinear integral equation valid for semi-infinite turbid media (Ambartsumian 1943):

$$R = R_{ss} + \frac{\omega_0(\xi L(\eta) + \eta L(\xi) + \xi \eta M(\xi, \eta))}{\xi + \eta}, \tag{3.79a}$$

where

$$L(\eta) = \frac{1}{4\pi} \int\limits_0^1 \int\limits_0^{2\pi} p(\eta, \varphi, \eta', \varphi') R(\eta', \varphi', \xi, \varphi_0) d\eta' d\varphi', \tag{3.79b}$$

$$L(\xi) = \frac{1}{4\pi} \int\limits_0^1 \int\limits_0^{2\pi} p(\xi, \varphi_0, \eta', \varphi') R(\eta', \varphi', \eta, \varphi) d\eta' d\varphi', \tag{3.79c}$$

$$M(\xi) = \frac{1}{4\pi^2} \int\limits_0^{2\pi} d\varphi' \int\limits_0^1 d\eta' R(\eta', \varphi', \eta, \varphi)$$

$$\times \int\limits_0^{2\pi} d\varphi'' \int\limits_0^1 d\eta'' p(-\eta', \varphi', \eta'', \varphi'') R(\eta'', \varphi'', \xi, \varphi_0). \tag{3.79d}$$

One can see that L and M depend on the reflection function R. There are different methods to solve the nonlinear integral equation described above. One possibility is to use the successive order approach, when R_{ms} is estimated from Eq. (3.79a) assuming that $R = R_{ss}$ in the first approximation. This enables the calculation of the double scattering approximation for the reflection function, which can be used to derive the triple scattering approximation, etc. Such series are converging quite fast at small values of ω_0. Otherwise, a lot of terms are required to reach a convergence. The functions L and M depend just on the phase function and the single scattering albedo and can be parameterized using numerical radiative transfer calculations of the reflection function.

Let us assume that isotropic scattering occurs in a nonabsorbing semi-infinite medium. Then both plane r_p and spherical r_s albedos are equal to 1.0. They are defined as

$$r_p(\xi) = 2 \int\limits_0^1 \overline{R}(\xi, \eta) \eta d\eta, \quad r_s = 2 \int\limits_0^1 r_p(\xi) d\xi, \tag{3.80}$$

where

$$\overline{R}(\xi, \eta) = \frac{1}{2\pi} \int\limits_0^{2\pi} R(\xi, \eta, \varphi) d\varphi. \tag{3.81}$$

It should be pointed out that the reflectance R is defined as the ratio of intensity of reflected light from a given target to the reflectance of absolutely white Lambertian surface. The isotropic scattering at single scattering albedo equal 1.0 (no absorption) is very close to the case of a Lambertian white surface. Therefore, one can assume that R is close to one. Then one can use the following approximation:

$$\int_0^1 \overline{R}(\xi, \eta) d\eta = 1. \tag{3.82}$$

This means that $L = 1/2$, $M = 1$ [see Eqs. (3.79)] and it follows for the isotropic scattering from Eq. (3.79) at the single scattering albedo $\omega_0 = 1$:

$$R_{ms} = \frac{\xi + \eta + 2\xi\eta}{2(\xi + \eta)}. \tag{3.83}$$

Therefore, one derives for the case of isotropic scattering at $\omega_0 = 1$:

$$R = \frac{1}{2} + \frac{\xi\eta}{\xi + \eta} + \frac{1}{4(\xi + \eta)}. \tag{3.84}$$

We assume that the dependence of L and M on the angles can be neglected in the first approximation also for other types of turbid media. Then it follows:

$$R_{ms} = \frac{\Gamma_1(\xi + \eta) + \Gamma_2\xi\eta + \Gamma_3}{2(\xi + \eta)} \tag{3.85}$$

and the constants Γ_j ($j = 1, 2, 3$) can be derived from fitting to this equation to the radiative transfer calculations. It follows from the discussion given above that $\Gamma_1 = \Gamma_2 = 1$, $\Gamma_3 = 0$ for isotropic scattering. It follows (Kokhanovsky 2005) for the fractal ice grain phase function: $\Gamma_1 = 0.593$, $\Gamma_2 = 2.579$, and $\Gamma_3 = 0.624$. Therefore, the following approximation for the nonabsorbing snow reflectance can be used:

$$R_0 = \frac{\Gamma_1(\xi + \eta) + \Gamma_2\xi\eta + \Gamma_3 + p(\theta)/2}{2(\xi + \eta)}, \tag{3.86}$$

where $p(\theta)$ is the snow phase function. One can use the parameterizations of the phase function as given by Eqs. (2.114) and (2.115). The reflection function of the semi-infinite turbid layer calculated using RTE (solid lines) and Eq. (3.86) (dashed line 1) for turbid media composed of nonabsorbing spherical water droplets (water cloud) and ice fractal particles (snow) as the function of the incidence angle at nadir observation and the wavelength 550 nm is shown in Fig. 3.1. The spherical water droplets are characterized by the effective droplet radius 6 microns and the coefficient of variance of gamma size distribution 38% (Deirmendjian 1969). The phase function of the fractal particles have been calculated using the fractal model of

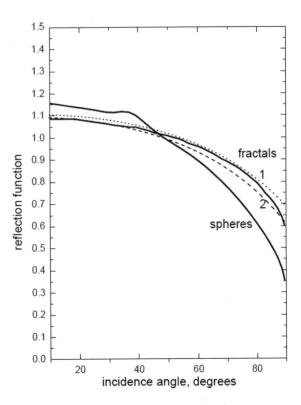

Fig. 3.1 The reflection function of the semi-infinite turbid layer calculated using RTE (solid lines) with the code developed by Mishchenko et al. (1999) and Eq. (3.86) (dashed line 1) for turbid media composed of nonabsorbing spherical water droplets (water cloud) and ice fractal particles (snow) as the function of the incidence angle at nadir observation and the wavelength 550 nm. The refractive index of water and ice was taken from Hale and Querry (1973) and Warren (1984), respectively. The dashed line 2 corresponds to Eq. (3.86) with $p(\theta) = 0$

ice grains (Macke and Tzschihholz 1992; Macke et al. 1996). It follows that Eq. (3.86) can be used to describe the angular behaviour of the snow reflection function at channels, where light absorption effects by snow can be neglected. One can see that the reflection function for the case of fractal snow grains is smooth and closer to the Lambertian case ($R = 1$) as compared to the water cloud model. Also there is no peak for the reflectance of turbid media with fractal particles in the rainbow region characteristic for water droplets. Therefore, one concludes that the assumption of spherical ice particles can lead to substantial errors in the interpretation of angular light intensity dependence for the case of light reflected from snow layers. This is not so crucial for the light transmitted by thick snow layers because of randomization of photon directions.

3.4.2 Reflection Function of Semi-Infinite Weakly Absorbing Snow Layers

Let us consider the case of weakly absorbing semi - infinite snowpack. The reflectance can be presented as (Rosenberg 1962; Kokhanovsky 2002):

$$R(\beta) = \sum_{n=1}^{\infty} a_n (1 - \beta)^n, \qquad (3.87)$$

where $\beta = 1 - \omega_0$ is the probability of photon absorption by unit volume of snow. In the case of nonabsorbing snow it follows:

$$R(0) = \sum_{n=1}^{\infty} a_n. \qquad (3.88)$$

Let us define the ratio:

$$\mathfrak{R} = R(\beta)/R(0). \qquad (3.89)$$

Then we have:

$$\mathfrak{R} = \frac{\sum\limits_{n=1}^{\infty} a_n (1 - \beta)^n}{\sum\limits_{n=1}^{\infty} a_n} \qquad (3.90)$$

Expanding the nominator of Eq. (3.88), we derive at small β:

$$\mathfrak{R} \approx 1 - \beta < n > + \frac{\beta^2}{2} < n^2 > - \frac{\beta^3}{6} < n^3 > + \ldots \approx < \exp(-\beta n) >, \qquad (3.91)$$

where

$$\langle n^k \rangle \equiv \sum_{n=1}^{\infty} f_n n^k, \quad \langle \exp(-\beta n) \rangle \equiv \sum_{n=1}^{\infty} f_n \exp(-\beta n), \quad f_n = \frac{a_n}{\sum\limits_{n=1}^{\infty} a_n} \qquad (3.92)$$

and we assumed that that $n(n - 1) \approx n^2$, $n(n - 1)(n - 2) \approx n^3 \ldots$ in our derivations. This is possible because β is close to zero and the number of scattering events is high. For the same reason we have:

$$< \exp(-\beta n) > \approx \int_0^{\infty} f(n) \exp(-\beta n) dn \qquad (3.93)$$

This integral can be evaluated assuming the function $f(n)$. In particular, it follows from the random walk theory that that the probability of a particle (photon) appearing at a given place, time, and direction after large number of iterations n can be presented as (Chandrasekhar 1943):

$$f(n) = \sqrt{\frac{\upsilon}{\pi}} n^{-3/2} \exp(-\upsilon/n). \tag{3.94}$$

The parameter υ depends on the process studied. The substitution of Eq. (3.94) into Eq. (3.93) gives:

$$\Re = \exp(-2\sqrt{\upsilon\beta}). \tag{3.95}$$

Therefore, we can write:

$$R(\beta) = R_0 \exp(-\sqrt{Q\beta}), \tag{3.96}$$

where $R_0 = R(0)$, $Q = 4\upsilon$. This equation has been proposed for the first time by Rosenberg (1962). It shows that the spectral snow reflectance depends on the square root of the probability of photon absorption β. The parameter s depends on the scattering and not on absorption processes and, therefore, one may assume that it does not depend on the wavelength for snow composed of large snow grains in contact. It should be pointed out that we have derived Eq. (3.96) not using the radiative transfer equation. Actually, the applicability of RTE for closely—packed media such as snow is in question. It should be pointed out that Eq. (3.79) is very general and can be applied to many geophysical media such as snow, leaves, etc. The main shortcoming of this equation as applied to natural snow is the fact that it has larger errors in areas close to forward scattering (glint) and backward scattering, where snow has enhanced brightness, e.g. due to oriented snow crystal surfaces or thin ice films on the top of snowpack. This equation can be applied only for the horizontal smooth snow surfaces. In particular, a modification is needed for a sloppy terrain (Picard et al. 2020). Zuravleva and Kokhanovsky (2011) has studied the influence of surface roughness on the radiation reflected from snow surfaces. It was found that the albedo of a snow layer is reduced (in particular, in the infrared region), if 3D effects are taken into account.

The value of Q can be related to the asymmetry parameter g of ice grains using asymptotic results of RTE valid at small values of the probability of photon absorption. Then it follows from Eq. (3.75):

$$R(\beta) = R_0(1 - \sqrt{Q\beta}) \tag{3.97}$$

and also from analytical radiative transfer theory (Sobolev 1975):

$$R(\beta) = R_0 - yu(\mu_0)u(\mu), \tag{3.98}$$

where

$$y = 4\sqrt{\frac{\beta}{3(1-g)}}. \qquad (3.99)$$

It follows that $y = 4s$ (see Table 2.1). The escape function can be found from the following equation (Sobolev 1975; Minin 1988):

$$u_0(\mu_0) = \frac{3}{4}(\mu_0 + \Psi(\mu_0)), \qquad (3.100)$$

where

$$\Psi(\mu_0) = 2\int_0^1 \overline{R}_0(\mu_0, \mu)\mu^2 d\mu, \ \overline{R}_0 = \frac{1}{2\pi}\int_0^1 R_0(\mu, \mu_0, \varphi)d\varphi. \qquad (3.101)$$

For weakly absorbing semi-infinite media one may expect that $\overline{R}_0$ is close to one and the dependence of Ψ on the angle can be neglected [see Eq. (3.100)]. Therefore, the escape function can be approximated by the linear function of μ_0.

The following normalization condition holds (Sobolev 1975):

$$2\int_0^1 u_0(\mu_0)\mu_0 d\mu_0 = 1. \qquad (3.102)$$

Also it follows for a special case of isotropic scattering (Sobolev 1975)

$$\int_0^1 u_0(\mu_0)d\mu_0 = 2u_0(0). \qquad (3.103)$$

Let us assume that the escape function can be represented as

$$u_0(\mu_0) = u_0(0)(1 + c\mu_0). \qquad (3.104)$$

One derives after substitution of Eq. (3.104) in Eq. (3.102), (3.103) that $c = 2$ and $u_0(0) = 3/7$ for the case of isotropic scattering. This means that

$$u_0(\mu_0) = \frac{3}{7}(1 + 2\mu_0) \qquad (3.105)$$

for the case of isotropic scattering. The escape function $u(\mu_0)$ has been tabulated by Yanovitskij (1997) for the Henyey-Greenstein phase function with the asymmetry parameter $g = 0.75$ close to that of snow. The results are given in Table 3.1. The data shown in Table 3.1 can be approximated by the following simple equation:

$$u_0(\mu_0) = \frac{3}{5}\mu_0 + \frac{1 + \sqrt{\mu_0}}{3}. \qquad (3.106)$$

Table 3.1 The escape function for the Henyey-Greenstein phase function at g = 0.75 (Yanovistskij 1997)

μ_0	0.0	0.1	0.2	0.3	0.4	0.5	0.6	0.7	0.8	0.9	1.0
u_0	0.333	0.484	0.595	0.692	0.783	0.869	0.952	1.033	1.113	1.192	1.27

Both Eqs. (3.105) and (3.106) satisfy the normalization condition (3.102). They have a similar accuracy at $\mu_0 > 0.2$. However, Eq. (3.106) gives better results at smaller values of μ_0 (see Fig. 3.2).

Comparing Eqs. (3.97) and (3.98), one derives:

$$Q = \frac{16f^2}{3(1-g)} \tag{3.107}$$

where

$$f = \frac{u(\mu_0)u(\mu)}{R_0}. \tag{3.108}$$

Therefore, we have finally (Zege et al. 1991):

$$R_\infty = R_{0\infty} \exp(-u(\mu)u(\mu_0)y/R_{0\infty}), \tag{3.109}$$

where we introduced the symbol ∞ to underline that we consider a semi-infinite turbid layer and y is given by Eq. (3.99). Essentially, Eq. (3.109) reduces the problem of the calculation of the reflection function of weakly absorbing turbid layer to that of a nonabsorbing layer ($R_{0\infty}$) with the same phase function. The reflection function

Fig. 3.2 The dependence of the escape function on the cosine of the solar zenith angle for the Henyey-Greenstein phase function at $g = 0.75$ calculated using exact radiative transfer theory (symbols), Eq. (3.105) (dashed line) and Eq. (3.106) (solid line)

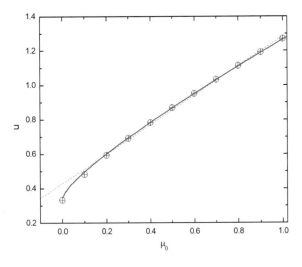

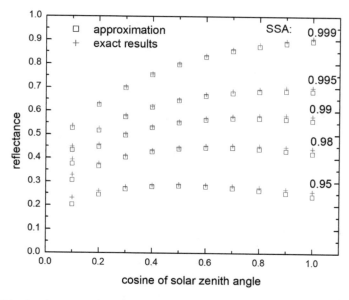

Fig. 3.3 The dependence of the reflectance of a semi-infinite snowpack on the cosine of the solar zenith angle at various values of single scattering albedo (SSA) and nadir observation. The Henyey-Greenstein phase at $g = 0.75$ has been used (exact results: crosses, approximation: boxes) (Kokhanovsky et al. 2019)

$R_{0\infty}$ depends only on the snow phase function and observation geometry. It can be easily parameterized as discussed in the previous section (see also Fig. 3.1).

The accuracy of Eq. (3.109) is demonstrated in Fig. 3.3 as compared with exact solution of RTE at the Henyey-Greenstein phase at $g = 0.75$ and various values of single scattering albedo (SSA). One can see that the analytical approximation is very accurate and can be used for the solution of many applied problems. The value of $R_{0\infty}$ has been calculated using the numerical solution of the radiative transfer equation. The error of the approximation is increased if an approximation (instead of exact value) for the function $R_{0\infty}$ is used. One can see that reflectance is smaller at the Sun located at the horizon (small values of cosine of incidence angle) as compared to the case of the illumination of snowpack along nadir direction (at the nadir observation). The reflectance of semi-infinite snow layers is in the range 0.2–0.3 at the single scattering albedo 0.95, which is quite low. This means that the probability of photon absorption larger than 0.05 leads to very small values of reflectance. Then the weak absorption approximation is not valid and snow is not a bright object anymore. The reflectance decreases with the solar zenith angle for nonabsorbing snow and also for a polluted snow with small values of β (see Fig. 3.3). For larger values of β, the dependence of reflectance on the solar zenith angle becomes nonmonotonic function with a maximum located around SZA = 60 degrees. This is due to the fact that photons have more chances to be absorbed in the medium in the case of nadir illumination. Therefore, the tendency of high light reflectance in the nadir observation direction characteristic for nonabsorbing media (as compared to the case of Sun at the horizon) is less pronounced or not valid anymore in the case larger values of probability of photon absorption (see Fig. 3.3).

3.4.3   Snow Albedo

Equation (3.109) can be used to find the plane albedo. For this, we shall use weak absorption limit of approximation (3.109):

$$R_\infty = R_{0\infty} - u(\mu)u(\mu_0)y \qquad (3.110)$$

Then it follows from Eq. (3.80), (3.103), (3.110):

$$r_{p\infty} = 1 - u(\mu_0)y, \qquad (3.111)$$

where we accounted for the fact that $r_{p\infty} = 1$ for nonabsorbing media by definition. We shall use the exponential approximation similar to Eq. (3.109) to extend the applicability of the derived approximation with respect to the value of the parameter y:

$$r_{p\infty} = \exp(-u(\mu_0)y). \qquad (3.112)$$

Using similar arguments we derive the following approximation for the snow spherical albedo:

$$r_{s\infty} = \exp(-y). \qquad (3.113)$$

Equation (3.113) can be also derived using Feynman path integrals technique (Perelman et al. 1994; Feynman and Hibbs 1965). In particular, it follows (Perelman et al. 1994):

$$r_{s\infty} = \exp\left[-2\sqrt{\frac{\Upsilon \ln \frac{1}{\omega_0}}{(1-g)}}\right], \qquad (3.114)$$

where Υ is the constant, which does not depend on ω_0. Equations (3.110) and (3.111) coincide at $\omega_0 \to 1$ and $\Upsilon = 4/3$.

It follows from Eqs. (3.106), (3.110) (Zege et al. 1991; Kokhanovsky and Zege 2004; Kokhanovsky et al. 2005a, 2007, 2018):

$$R_\infty = R_{0\infty} r_{s\infty}^f, \qquad (3.115)$$

$$r_{p\infty} = r_{s\infty}^{u(\mu_0)}.$$ (3.116)

It follows from Eq. (3.116):

$$r_{p\infty}(\mu_{01}) = r_{p\infty}^{\kappa}(\mu_{02}),$$ (3.117)

where $\kappa = u(\mu_{01})/u(\mu_{02})$, μ_{01} is the cosine of the solar zenith angle ϑ_{01}, μ_{02} is the cosine of the solar zenith angle ϑ_{02}. This equation makes it possible to determine the snow plane albedo at any solar zenith angle, if the plane albedo is known (measured) at only one solar zenith angle (under the assumption that snow microstructure is not changing during the day).

Equation (3.115) can be used for the determination of spherical albedo from the snow reflection function measurements:

$$r_{s\infty} = (R_\infty/R_{0\infty})^{1/f}.$$ (3.118)

Then the plane albedo can be derived from Eq. (3.116). The accuracy of approximations for the plane and spherical albedos is shown in Figs. 3.4 and 3.5.

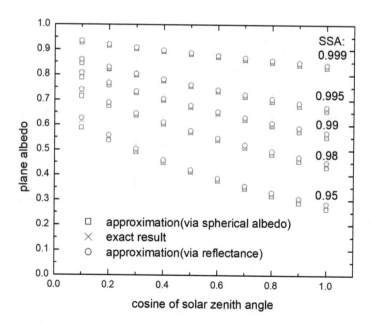

Fig. 3.4 The plane albedo as the function of the cosine of the solar zenith angle at various values of single scattering albedo for the semi-infinite snow layer with the asymmetry parameter $g = 0.75$. The exact results are presented together with two approximations [boxes-Eq. (3.92), circles-calculation of plane albedo using Eq. (3.89)] (Kokhanovsky et al. (2020)

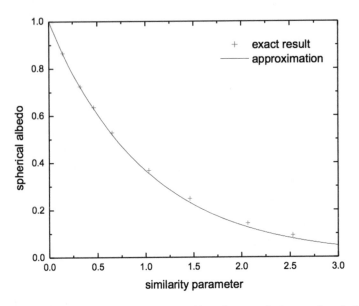

Fig. 3.5 The spherical albedo derived using Eq. (3.90) and exact radiative transfer calculations as function of the parameter $y = 4s$ (Kokhanovsky et al. 2020)

The spherical albedo $r_{s\infty}$ can be used to determine the semi-infinite snow absorptance under diffuse illumination conditions: $\Pi = 1 - r_{s\infty}$. It follows in the limit of weak absorption: $\Pi = y$. This makes it possible to establish the physical meaning of the parameter y. Namely, this parameter provides the value of snow absorptance at $\beta \to 0$.

The parameter y can be related to the size of ice grains in snow. In particular, it follows as $\beta \to 0$ [see Eq. (2.109)]:

$$y = \sqrt{\alpha \ell}, \qquad (3.119)$$

where $\ell = \varepsilon d_{ef}$ is the effective absorption length (EAL), which is proportional to the snow grain effective diameter, and

$$\varepsilon = \frac{16B}{9(1 - g)}. \qquad (3.120)$$

The parameter ε depends on the shape of particles. However, its dependence on the wavelength and the size of particles in the spectral range 0.4–1 μm can be neglected (see the results presented in the previous Chapter). Therefore, we conclude that the EAL can be derived from spectral snow reflectance measurements even, if the shape of snow crystals is not known in advance. The determination of the effective grain diameter requires the parameter ε, which depends on the shape of ice crystals/snow type.

It follows from Eqs. (3.113), (3.119):

$$r_{s\infty}(\lambda) = \exp(-\sqrt{\alpha(\lambda)\ell}), \qquad (3.121)$$

where $\alpha(\lambda) = \frac{4\pi\chi(\lambda)}{\lambda}$ is the bulk ice absorption coefficient given in Appendix. Because the function $\alpha(\lambda)$ is well known, we can conclude that the spectral diffuse reflection (under diffuse illumination conditions) of a flat clean snow layer depends on just one parameter, namely-the effective absorption length L, which can be derived from the measurements of the spherical albedo at a single wavelength in the near infrared (say, around 1 μm). The same is true for the plane albedo [see Eq. (3.112)]. The accuracy of this approximation is demonstrated in Fig. 3.6. It follows that the proposed approximation can be used in the spectral range 300–1200 nm with a high accuracy. The accuracy decreases at larger wavelengths due to the fact that the error of the weak absorption approximation increases and also due to the fact that one must account for the different penetration depths of radiation to different layers of vertically inhomogeneous snow. Equation (3.121) does not take into account neither surface roughness no vertical snow inhomogeneity effects. The slight discrepancy of measurements and theory around 400 nm can occur due to the fact that our theoretical model ignores possible contamination of snow by various impurities. The uncertainty in the imaginary part of the complex ice refractive index (see Appendix) can play a role as well.

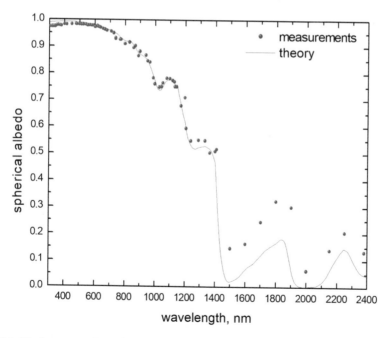

Fig. 3.6 The intercomparison of calculations using Eq. (3.121) (line) with experimental measurements (symbols) performed in Antarctica (Grenfel et al. 1994). The experimental results are derived using averaging of data (19 spectra) for two stations (Vostok and South Pole)

3.4.4   Snow Broadband Albedo

In some cases the broadband albedo (BBA) is of interest. It is defined as

$$r_b = \frac{\int\limits_{\lambda_1}^{\lambda_2} r(\lambda) F(\lambda) d\lambda}{\int\limits_{\lambda_1}^{\lambda_2} F(\lambda) d\lambda}, \tag{3.122}$$

where $F(\lambda)$ is the solar flux incident on the snow surface and $r(\lambda)$ is either spherical or plane snow albedo depending on the illumination conditions. Clearly, for a cloudy sky, $r(\lambda)$ in Eq. (3.122) is the spherical albedo. For a cloudless clean atmosphere, $r(\lambda)$ is the plane albedo. We show the spectral solar flux $F(\lambda)$ for cloudless atmosphere at various solar zenith anlges (SZA) in Fig. 3.7. The results have been obtained using the SBDART (Ricchiazi et al. 1998) radiative transfer model. The absorption features of various atmospheric gases are clearly seen in this figure. The parameters of calculations using SBDART radiative transfer model were:

- the water vapour column: 2.085 g/m^2,
- the ozone column: 0.35 atm-cm,
- the tropospheric ozone: 0.0346 atm-cm,

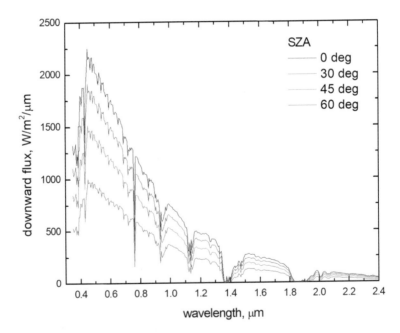

Fig. 3.7 The solar flux at the snow surface (courtesy: M. Lamare)

Table 3.2 The coefficients of approximation given by Eq. (3.123)

p_0	p_1	p_2	λ_1, microns	λ_2, microns
3.238e + 1	−1.6014033e + 5	7.95953e + 3	8.534e−4	4.0179e−1

- the aerosol model: rural (Shettle and Fen 1979),
- the vertical optical depth of boundary layer at 550 nm: 0.1,
- the altitude: 825 m,
- the snow albedo at the surface calculated using the assumption that the grain diameter is equal to 0.25 mm and grains are of spherical shape.

Taking into account that the downward solar flux appears in the nominator and the dominator of Eq. (3.119), the smale-scale oscillations of the curve for the calculation of the flux is of a secondary importance. Therefore, for the back-of-envelope calculations one may use the approximation of the downward flux given below:

$$F(\lambda) = p_0 + p_1 e^{-\lambda/\lambda_1} + p_2 e^{-\lambda/\lambda_2}, \tag{3.123}$$

where the oscillations of the downward flux are ignored and the coefficients derived at SZA $= 60$ degrees are presented in Table 3.2.

The limits λ_1 and λ_2 in Eq. (3.122) can vary. In particular, CMP21 (Kipp& Zonen, Delft, Netherlands) pyranometers measure.

- near IR BBA ($\lambda_1 = 708$ nm, $\lambda_2 = 2800$ nm);
- shortwave BBA ($\lambda_1 = 305$ nm, $\lambda_2 = 2400$ nm).

The visible BBA ($\lambda_1 = 305$ nm, $\lambda_2 = 708$ nm) is obtained by subtracting the near IR BBA from the shortwave BBA.

The spherical BBA can be derived substituting Eq. (3.118) into Eq. (3.119). In particular, it follows:

$$r_b = r_s(\lambda^*) = \exp(-\sqrt{\alpha(\lambda^*)L}), \tag{3.124}$$

where we used the mean value theorem for the evaluation of the integral in the nominator. Clearly, the wavelength λ^* depends on the value of L. The similar result holds for the polluted snow as well.

Accurate estimations of the BBA require the numerical solution of the radiative transfer equation (see, e.g., Aoki et al. (2011), who also proposed the account for the partial illumination of a snow surface by diffuse and direct light). In particular, Aoki et al. (2011) has proposed the following expression for the shortwave BBA:

$$r_{b,sw} = \frac{F_{vis} r_{b,vis} + F_{NIR} r_{b,NIR}}{F_{vis} + F_{NIR}}, \tag{3.125}$$

where F_{vis} are global solar radiation in the visible, F_{NIR} is the global solar radiation in the NIR,

$$r_{b,vis} = (1 - f_{vis})r_{b,p}^{vis} + f_{vis}r_{b,s}^{vis}, \tag{3.126}$$

$$r_{b,NIR} = (1 - f_{NIR})r_{b,p}^{NIR} + f_{NIR}r_{b,s}^{NIR}. \tag{3.127}$$

f_{vis} (f_{NIR}) is the diffuse fraction of the global solar radiation in the visible (NIR), $r_{b,p}^{vis}$ is the visible plane BBA, $r_{b,s}^{vis}$ is the visible spherical BBA, $r_{b,p}^{NIR}$ is the NIR plane BBA, $r_{b,s}^{NIR}$ is the NIR spherical BBA. Sometimes the spherical (plane) albedo is referenced as white (black) sky albedo.

The broadband albedo of snow is of importance for climate studies and also for general circulation models. It is measured at many points worldwide including the **PRO**gram for Monitoring of the Greenland **ICE** Sheet (**PROMICE**) network of weather stations. The locations of stations are shown by red points in Fig. 3.8. The BBA is measured at all illumination conditions (clean, cloudy sky, see Figs. 3.8, 3.9).

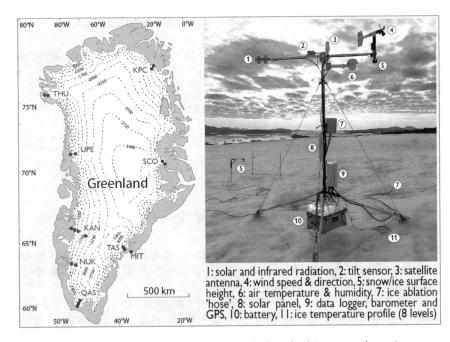

1: solar and infrared radiation, 2: tilt sensor, 3: satellite antenna, 4: wind speed & direction, 5: snow/ice surface height, 6: air temperature & humidity, 7: ice ablation 'hose', 8: solar panel, 9: data logger, barometer and GPS, 10: battery, 11: ice temperature profile (8 levels)

Fig. 3.8 The PROMICE network of weather stations in Greenland (www.promice.org)

Fig. 3.9 The automatic measurements of snow BBA at the KAN_U PROMICE station (April 12, 2013) (courtesy: J. P. Box)

3.5 Finite Optically Thick Snow Layers: Reflection and Transmission

3.5.1 Ambartsumian Approximation

3.5.1.1 General Equations

Let us assume now that the snow optical thickness (SOT) is such that snowpack can not be treated as a semi-infinite layer. However, it is supposed that SOT is large (above 5–10). Then simple approximations for light reflection and transmission by a snowpack can be derived. Let us show it using the Ambartsumian approximation (Ambartsumian 1944, 2011) for the diffuse transmittance t and the diffuse reflectance (spherical albedo) r of a finite turbid layer. To simplify, we study the radiative transfer along a line with the optical thickness τ (an idealized 1-D medium). This medium reflects some fraction $r I_0$ of initial intensity I_0 and transfers a fraction $t I_0$ along a line.

Now we assume that this 1-D medium is composed of two 1-D media: one with optical thickness τ_1 and yet another with the optical thickness τ_2. Clearly, the transmitted I_t and reflected I_r light intensities will given by the following equations:

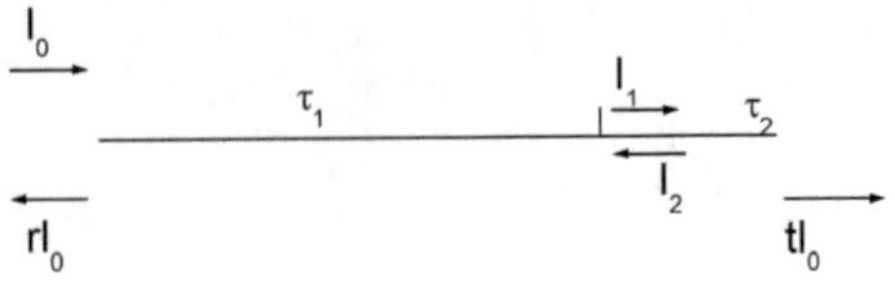

Fig. 3.10 The radiative
transfer in 1-D medium

$$I_t = t(\tau_1 + \tau_2)I_0, \quad I_r = r(\tau_1 + \tau_2)I_0. \tag{3.128}$$

We shall also introduce light intensities *at the boundary* between two media (see Fig. 3.10) and call them I_1 (for the radiation in the direction of incidence after propagating the layer of optical thickness τ_1) and I_2 (for the radiation propagating in the direction opposite to the incidence direction after reflection from the layer of optical thickness τ_2).

Clearly, it follows:

$$I_2 = r(\tau_2)I_1 \tag{3.129}$$

and, on the other hand,

$$I_1 = t(\tau_1)I_0 + r(\tau_1)I_2. \tag{3.130}$$

Then one derives from Eqs. (3.129), (3.130):

$$I_1 = t(\tau_1)I_0 + r(\tau_1)r(\tau_2)I_1. \tag{3.131}$$

This means that

$$I_1 = \frac{t(\tau_1)}{1 - r(\tau_1)r(\tau_2)} I_0 \tag{3.132}$$

and

$$I_t = \frac{t(\tau_1)t(\tau_2)}{1 - r(\tau_1)r(\tau_2)} I_0 \tag{3.133}$$

and, therefore,

$$t(\tau_1 + \tau_2) = \frac{t(\tau_1)t(\tau_2)}{1 - r(\tau_1)r(\tau_2)}. \tag{3.134}$$

For the reflected radiation, we may write:

$$I_r = t(\tau_1)I_2 + r(\tau_1)I_0. \tag{3.135}$$

Therefore, it follows:

$$r(\tau_1 + \tau_2) = r(\tau_1) + \frac{r(\tau_2)t^2(\tau_1)}{1 - r(\tau_1)r(\tau_2)}. \tag{3.136}$$

The derived Eqs. (3.134), (3.136) show how the reflectance and transmittance of a combined layer can be expressed via reflectance and transmittance of two individual layers. These two sub-layers could have not only different optical thicknesses but also they can have different absorption and scattering properties. In a similar way, we may consider an arbitrary number of layers. In the discussion given below we assume that both media have the same scattering and absorption properties.

3.5.1.2 Nonabsorbing Media

Let assume that there is no absorption in the medium. Then it follows: $r(\tau) + t(\tau) = 1$ and, therefore,

$$t(\tau_1 + \tau_2) = \frac{t(\tau_1)t(\tau_2)}{t(\tau_1) + t(\tau_2) - t(\tau_1)t(\tau_2)} \tag{3.137}$$

or taking inverse values:

$$t^{-1}(\tau_1 + \tau_2) = t^{-1}(\tau_1) + t^{-1}(\tau_2) - 1. \tag{3.138}$$

We may also write this equation in the following form:

$$t^{-1}(\tau_1 + \tau_2) - 1 = t^{-1}(\tau_1) - 1 + t^{-1}(\tau_2) - 1. \tag{3.139}$$

Let us introduce the function

$$F(\tau) = t^{-1}(\tau) - 1. \tag{3.140}$$

One can see that this function has the following property:

$$F(\tau_1 + \tau_2) = F(\tau_1) + F(\tau_2) \tag{3.141}$$

and, therefore,

$$F(\tau) = c\tau, \tag{3.142}$$

where c does not depend on τ. Clearly, this parameter must depend on the peculiarities of light scattering (isotropic/anisotropic light scattering, etc.) in the layer under study. An important point is that we arrive to the following equation:

$$t^{-1}(\tau) = 1 + c\tau \qquad (3.143)$$

and, therefore, it follows for the transmission and reflection coefficients of a turbid layer:

$$t(\tau) = \frac{1}{1 + c\tau}, r(\tau) = \frac{c\tau}{1 + c\tau}. \qquad (3.144)$$

The numerical calculations using the radiative transfer equation and also asymptotic analysis of the RTE (Sobolev 1975; Minin 1988; van de Hulst 1980) show that the inverse transmittance can be actually presented as

$$t^{-1}(\tau) = b + c\tau, \qquad (3.145)$$

where 1.072 and $c = 0.75(1 - g)$ independently on the actual phase function. Therefore, Eqs. (3.144) must be substituted by more accurate equations:

$$t(\tau) = \frac{1}{b + c\tau}, r(\tau) = \frac{c\tau}{b + c\tau}. \qquad (3.146)$$

These equations are of great importance for the snow optics because they show how the diffuse reflection and transmission of a nonabsorbing snow layer (say, in the visible for clean snow) changes with optical thickness (assuming black underlying surface), if the semi-infinite layer approximation does not work.

3.5.1.3   Absorbing Media

Now we consider the case, when one can not neglect light absorption processes in an optically thick but a finite layer. Again we concentrate on the 1-D case. Let us assume that the incident light intensity is I_0. Then the element of optical thickness $d\tau$ absorbs the following fraction of light:

$$a = (1 - \omega_0) I_0 d\tau. \qquad (3.147)$$

The reflected light fraction will be:

$$r = (1 - f) \omega_0 I_0 d\tau, \qquad (3.148)$$

where f describes the probability of scattering in a given direction. It is equal to f in the forward scattering direction and $(1 - f)$ in the backward scattering direction for 1-D case. Clearly, we should have due to the conservation of energy law:

$$r + t + a = 1. \tag{3.149}$$

Therefore, it follows for the transmission:

$$t = 1 - (1 - \omega_0 f) d\tau. \tag{3.150}$$

Now we apply equations given above at $\tau_1 = \tau, d\tau = \tau_2$:

$$r(\tau + d\tau) = r(\tau) + \frac{(1 - f)\omega_0 t^2(\tau) d\tau}{1 - r(\tau)(1 - f)\omega_0 d\tau}, \tag{3.151}$$

$$t(\tau + d\tau) = \frac{t(\tau)(1 - (1 - \omega_0 f) d\tau}{1 - r(\tau)(1 - f)\omega_0 d\tau}. \tag{3.152}$$

Let us use the linear approximation for differences $dr(\tau) = r(\tau + d\tau) - r(\tau)$ and $dt(\tau) = t(\tau + d\tau) - t(\tau)$ with respect to $d\tau$. Then it follows:

$$\frac{dr(\tau)}{d\tau} = (1 - f)\omega_0 t^2(\tau), \tag{3.153}$$

$$\frac{dt(\tau)}{d\tau} = (1 - f)\omega_0 r(\tau) t(\tau) - (1 - \omega_0 f) t(\tau). \tag{3.154}$$

The last equation takes the following form at $\omega_0 = 1$:

$$\frac{dt(\tau)}{d\tau} = -(1 - f) t^2(\tau). \tag{3.155}$$

with a solution given by Eq. (3.144) at $c = 1 - f$. One can see that the derivative of the diffuse transmittance at $\omega_0 = 1$ is proportional to the value of t^2 and not to t as it the case for the direct light.

The system of two simultaneous ordinary differential Eqs. (3.153), (3.154) can be solved analytically. This makes it possible to derive analytical equations both for r and t. As a matter of fact, the system can be reduced to a single differential equation taking into account that the value of $t^2(\tau)$ in the first equation can be presented as

$$t^2(\tau) = 1 + r^2(\tau) - br, \tag{3.156}$$

where

$$b = \frac{2(1 - \omega_0 f)}{\omega_0(1 - f)}. \tag{3.157}$$

This equation can be derived by division of the first equation by the second one and subsequent integration. Namely, it follows:

$$t\,dt = r\,dr - \frac{b}{2}dr \tag{3.158}$$

or

$$t^2 = r^2 - br + C, \tag{3.159}$$

where the integration constant $C = 1$ because $t \equiv 1$ and $r = 0$ at $\tau = 0$. Therefore, it follows from Eq. (3.153):

$$\frac{dr(\tau)}{\varepsilon d\tau} - r^2(\tau) + br(\tau) - 1 = 0, \tag{3.160}$$

where

$$\varepsilon = (1 - f)\omega_0. \tag{3.161}$$

This nonlinear differential equation can be solved analytically. Let us introduce the new coordinate: $z = (1 - f)\omega_0\tau$. Then the equation to be solved can be presented in the following form:

$$\frac{dr(z)}{dz} = (r(z) - r_0^{-1})(r(z) - r_0), \tag{3.162}$$

where r_0^{-1} and r_0 are the roots of the polynomial

$$\psi(z) = 1 + r^2(z) - br(z). \tag{3.163}$$

This differential equation has the following solution:

$$r = r_0 \frac{1 - \exp(-\nu z)}{1 - r_0^2 \exp(-\nu z)}, \tag{3.164}$$

where $\nu = (r_0^{-1} - r_0)$. Let us return to the initial coordinate τ. Then we derive:

$$r = r_0 \frac{1 - \exp(-2k\tau)}{1 - r_0^2 \exp(-2k\tau)}, \tag{3.165}$$

where

$$k = \frac{1}{2}(1 - f)(r_0^{-1} - r_0)\omega_0. \tag{3.166}$$

We derive for the transmittance from Eq. (3.116):

$$t = \frac{(1 - r_0^2)\exp(-k\tau)}{1 - r_0^2\exp(-2k\tau)}. \tag{3.167}$$

One can see from Eq. (3.122) that r_0 is equal to the value of r for a semi-infinite layer. Therefore, it follows that $r_0 = e^{-y}$ as discussed above. Let us introduce the parameter $x = k\tau$ or $x = \gamma L$, where L is the snow geometrical thickness and $\gamma = kk_{ext}$ is the asymptotic flux attenuation coefficient. The coefficient k is called the diffusion exponent.

Then we derive:

$$r = \frac{\exp(-y)(1 - \exp(-2k\tau))}{1 - \exp(-2k\tau - 2y)}, t = \frac{(1 - \exp(-2y))\exp(-x)}{1 - \exp(-2y - 2x)} \tag{3.168}$$

or alternatively

$$r = \frac{shx}{sh(x + y)}, \tag{3.169}$$

$$t = \frac{shy}{sh(x + y)}. \tag{3.170}$$

Assuming that $\omega_0 \to 1$, one derives from Eq. (3.166):

$$k = 4(1 - f)\sqrt{\frac{1 - \omega_0}{3(1 - g)}}. \tag{3.171}$$

The asymptotic analysis of RTE shows that in reality [see Eq. (3.62)]

$$k = \sqrt{3(1 - g)(1 - \omega_0)}, \tag{3.172}$$

as $\omega_0 \to 1$, which means that the empirical parameter $\epsilon = 1 - f$ given in approximate equations presented above must be substituted by $\epsilon = 0.75(1 - g)$. One can see that reflection and transmission characteristics of a turbid layer with a given optical thickness for weakly absorbing media depend on just two parameters (k, $y = 4$ s) as it was in the case of asymptotical regime inside turbid layers far from its boundaries. The first parameter (k) describes the rate of attenuation of light field in deep layers

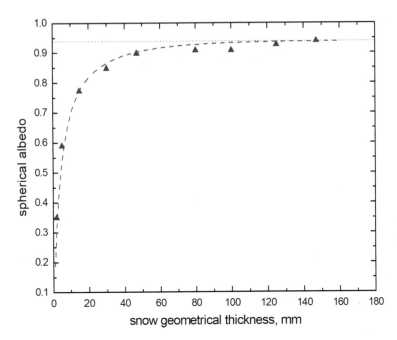

Fig. 3.11 The dependence of the spherical albedo on the snow geometrical thickness at the wavelength 450 nm [symbols—measurements (Perovich 2007), line—Eq. (3.169)]

of turbid media. The second parameter (y) describes the light reflection from a semi-infinite layer of a turbid medium. Various turbid media of a given optical thickness τ having similar values (k, y) also have similar optical properties. Equation (3.144) follows from Eqs. (3.169), (3.170) at probability of photon absorption equal to zero. More accurate Eq. (3.146) can be derived from Eqs. (3.169), (3.170), if the value of y in the dominator of these equations is substituted by the parameter $Y = by$.

The spherical albedo r calculated using Eq. (3.169) is shown in Fig. 3.11 as function of the snow geometrical thickness L. Also we show the measured spherical albedo for the natural snow cover over various thicknesses (Perovich 2007) on the same figure. The value of y in Eq. (3.169) has been derived as

$$y = \ln\left(\frac{1}{r_\infty}\right), \tag{3.173}$$

where r_∞ is the measured spherical albedo at the largest snow geometrical thickness. The asymptotic flux attenuation coefficient has been derived using the following equation:

$$\gamma = \frac{1}{2L} \ln\left\{\frac{1 - r^* r_\infty}{1 - r^*/r_\infty}\right\}, \tag{3.174}$$

which follows from Eq. (3.169). Here r^* is the spherical albedo at the snow geometrical thickness L taken to be equal 15 mm for the experimental data shown in Fig. 3.11.

The value of x in Eq. (3.169) has been calculated as:

$$x = \gamma L. \tag{3.175}$$

One can see that Eq. (3.169) can indeed be used to describe the experimentally measured values of spherical albedo for finite snow layers. It has appeared that the value of AFEC is equal to $0.1464\ cm^{-1}$ for the case shown in Fig. 3.11. The e-folding scale is defined as $\zeta = 1/\gamma$. Therefore, it follows: $\zeta = 6.8$ cm. The value of ζ gives the distance, where the diffuse radiation attenuates in e—times for a given snowpack. The transport optical thickness van be derived from the values of $x = \gamma L$ and y. Namely it follows:

$$\tau_{tr} = \frac{4x}{3y}. \tag{3.176}$$

The transport extinction coefficient is derived as: $k_{xr} = \tau_{tr}/L$. Also one can determine the transport length: $l_{tr} = 1/\sigma_{tr}$. The transport length is the length over which the direction of propagation of photon is randomized (see Fig. 3.12). The mean free path is determined by the inverse value of the extinction coefficient $l_{mfp} = 1/k_{ext}$. Then it follows: $l_{tr} = l_{mfp}/(1-g)$. It follows that for the isotropic scattering both characteristic lengths (l_{mfp}, l_{tr}) coincide. For anisotropic scattering media such as snow $l_{tr} > l_{mfp}$. We shall assume that the snow asymmetry parameter in the visible and near infrared is equal to ¾ as for crystalline clouds. Then it follows: $l_{tr} = 4l_{mfp}$. Therefore, the direction of propagation is randomized along the distance equal to the 4 mean free path lengths in the snowpack.

Also the snow optical thickness can be derived assuming that the asymmetry parameter is known. Assuming that $g = 3/4$, one derives:

$$\tau = \frac{16x}{3y}. \tag{3.177}$$

Fig. 3.12 The transport length in a highly scattering medium

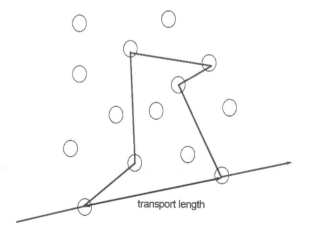

transport length

The transport extinction coefficient can be derived as $k_{tr} = \frac{\tau_{tr}}{L}$. It has appeared that $k_{tr} = 3.09$ cm^{-1} and, therefore, extinction coefficient $k_{ext} = k_{tr}/(1-g)$ is equal to 12.4 cm^{-1} for the case shown in Fig. 3.11. The extinction coefficient is related to the effective diameter of snow grains d_{ef}:

$$k_{ext} = \frac{3c}{d_{ef}}, \qquad (3.178)$$

where $c = \rho_s/\rho_i$, ρ_s is the snow density, ρ_i is the ice density. Taking into account that $\rho_s = 0.16$ gcm^{-3} for the experiment shown in Fig. 3.11, we derive from Eq. (3.167): $d = 0.42$ mm, which is close to estimates of the size grains given by Perovich (2007). This confirms the fact that the theory presented here can be used to describe the spherical albedo of finite snow layers. It follows from Fig. 3.11 that the limit of a semi-infinite snow is reached at the depth of approximately equal to 15 cm at 450 nm for the polluted snow studied by Perovich (2007). This means that adding additional snow layers under 15 cm thick snow layer will not change snow reflectance. Therefore, this thickness can be considered as the radiation penetration depth for the case studied. Although the contribution of the radiation from the depths larger than 5 cm is already small (see Fig. 3.11). Perovich (2007) has stated that the snow with the depth 8 cm over the measurement platform (painted black) appears as white as the surrounding deeper snow.

The theory presented above makes it possible to derive many important snow characteristics such as AFEC, transport extinction coefficient, extinction coefficient, snow optical thickness, and the snow grain size. The snow absorption optical thickness $\tau_{abs} = \tau_{ext} L$ can be also derived:

$$\tau_{abs} = \frac{xy}{4}. \qquad (3.179)$$

We show the spherical albedo calculated at various values of diffusion exponent k in Fig. 3.13. The experimental results are also given there. The snow optical thickness can be presented as

$$\tau = A\frac{L}{d}, \qquad (3.180)$$

where

$$A = 3\frac{\rho_s}{\rho_i}, \qquad (3.181)$$

$\rho_i = 0.917$ gcm^{-3} is the bulk ice density and ρ_s is the snow density. The value of ρ_s was equal to 0.16 gcm^{-1} for the experiment shown in Fig. 3.11. Therefore, it follows: $A = 0.522$. We have also estimated the ice grain diameter from the measurements as

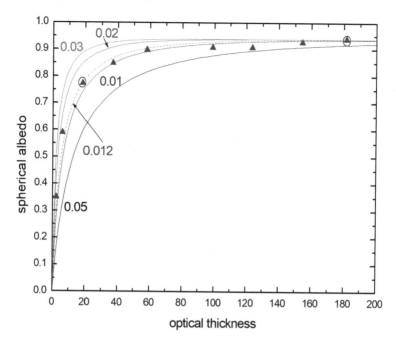

Fig. 3.13 The dependence of the spherical albedo on the snow optical thickness at the wavelength 450 nm and various values of the diffusion constant k [symbols—measurements (Perovich 2007), lines—Eq. (3.169)]. The fitting (see the dashed line) experimental results at two points shown by circles leads to the estimated value of *the diffusion exponent* equal to 0.012

discussed above. It appears that *the effective grain diameter is equal to* 0.42 mm. Therefore, it follows: $\tau = \zeta L$, where the scaling constant $\zeta = 1.236$. We have used this scaling constant to present experimental measurements of the spherical albedo as the function of snow optical thickness in Fig. 3.13. The deviation of experimental data from the dashed line shown in Fig. 3.11 is probably due to vertical variation of snow properties and also due to the errors of the measurements (albedo, snow geometrical thickness). It should be pointed that the snow optical thickness at which the limit of a semi-infinite snow is reached depends on the diffusion exponent k being smaller at larger values of k. One can state that the semi-infinite medium case is reached at SOT = 120 for the experiment shown in Fig. 3.11. This limiting value of SOT is much smaller in the near IR part of the electromagnetic spectrum.

3.5.2   Reflection and Transmission Functions of Nonabsorbing Snow Layers

Let us consider now the transmission function for an optically thick snow layer. Because $\tau \gg 1$, the dependence of transmittance on the azimuth can be neglected.

Also both theory and observations show that the angular distribution of the transmitted light can be presented as a linear function of the cosine of the observation angle. It is possible to show that the normalized angular distribution of the transmitted light (to its value in zenith direction) does not depend on the value of τ. Taking into account that we consider homogeneous layers, the principle of reciprocity must be satisfied. This means that

$$T_0(\xi, \eta, \tau) = t_0(\tau)u_0(\xi)u_0(\eta). \tag{3.182}$$

The separation of the angular and geometrical coordinates is one of main results valid for optically thick scattering layers.

The diffuse transmittance is defined as (for any value of single scattering albedo)

$$t = 2 \int_0^1 t_p(\xi)d\xi, \tag{3.183}$$

$$t_p(\xi) = 2 \int_0^1 \overline{T}(\xi, \eta)\eta d\eta, \tag{3.184}$$

where

$$\overline{T}(\xi, \eta) = \frac{1}{2\pi} \int_0^{2\pi} T(\xi, \eta, \varphi)d\varphi. \tag{3.185}$$

Taking into account the considerations presented in the previous section, we derive:

$$T_0(\xi, \eta, \tau) = \frac{1}{b + p\tau_{tr}}u_0(\xi)u_0(\eta), \tag{3.186}$$

where $b = 1.072$, $p = 0.75$ and $\tau_{tr} = k_{tr}l$ is so-called snow scaled or transport optical thickness, l is the geometrical thickness of snow layer and $k_{tr} = (1 - g)k_{ext}$ is the transport extinction coefficient. This equation can be used to study transmission function of a finite nonabsorbing snow layer. To derive k_{tr}, one needs to measure the transmittance and snow geometrical thickness L:

$$\sigma_{tr} = \frac{\tau_{tr}}{L}, \tag{3.187}$$

where

$$\tau_{tr} = \frac{4}{3}(T_0^{-1}(\xi, \eta, \tau)u_0(\xi)u_0(\eta) - \alpha) \tag{3.188}$$

and we assume black underlying surface. Similar equations can be derived if not the transmission function but rather the plane transmittance $t_p(\xi)$ or diffuse transmittance t are measured.

The accuracy of Eq. (3.186) as compared to the solution of RTE is given in Fig. 3.14. It follows that the transmission function depends on a single parameter-scaled optical thickness till $\tau_{tr} = 1$. The linear dependence of $1/T$ on the scaled optical thickness is valid till $\tau_{tr} = 1.5$ for the case studied.

Let us consider now the reflection function $R(\xi, \eta, \tau)$ assuming that there is no light absorption in the medium. Approximate equation for this function can be derived in the following way.

Let us take the semi-infinite nonabsorbing medium and make an artificial cut inside the medium at large optical thickness τ. Then the reflection function of a semi-infinite layer can be presented as the sum of the reflection function of optically thick upper layer and also the transmission function of radiation diffusely reflected from the lower semi-infinite nonabsorbing layer. Taking onto account that the diffuse reflection coefficient of the lower layer is 1, we derive:

$$R_{\infty 0}(\xi, \eta, \varphi, \tau) = R_0(\xi, \eta, \tau) + T_0(\xi, \eta, \tau). \qquad (3.189)$$

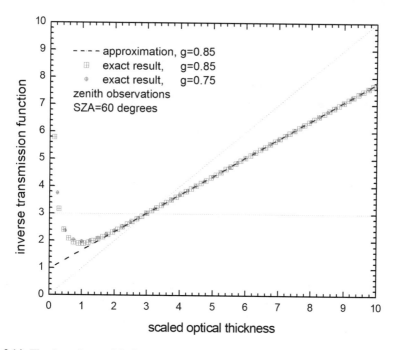

Fig. 3.14 The dependence of the inverse value of the transmission function on the scaled optical thickness calculated using Eqs. (3.186) and the numerical RTE solution at single scattering albedo $\omega_0 = 1$, the Henyey-Greenstein phase function and two values of the asymmetry parameter (0.75, 0.85) at the zenith direction. The solar zenith angle is equal to $60°$.

This means that

$$R_0(\xi, \eta, \tau) = R_{\infty 0}(\xi, \eta, \varphi) - T_0(\xi, \eta, \tau). \tag{3.190}$$

Therefore, the calculation of $R(\xi, \eta, \tau)$ for the case of nonabsorbing medium is reduced to the calculation of the functions studied above. Because the dependence of $R_{\infty 0}(\xi, \eta, \varphi, \tau)$ on the phase function can be neglected, one can use the parameterization to determine $R_{\infty 0}(\xi, \eta, \varphi)$ for a given geometry and then derive $T_0(\xi, \eta, \tau)$ and transport optical thickness from the reflectance measurements of a given finite snow layer.

3.5.3 Reflection and Transmission Functions of Weakly Absorbing Snow Layers

In the case of weakly absorbing layers, one can use similar equation as given above. However, the diffuse transmittance coefficient must be substituted by that presented. Then it follows (Zege et al. 1991):

$$T(\xi, \eta, \tau) = \frac{shy}{sh(x + by)} u_0(\xi) u_0(\eta), \tag{3.191}$$

where we have accounted for the fact that b differs from unity. The equation for the reflectance of a finite weakly absorbing layer can be derived in a similar way as discussed above. Then it follows:

$$R_\infty(\xi, \eta, \varphi) = R(\xi, \eta, \tau) + T(\xi, \eta, \tau) \exp(-x - y). \tag{3.192}$$

Therefore, one derives:

$$R(\xi, \eta, \varphi, \tau) = R_{\infty 0}(\xi, \eta, \varphi) \exp(-uy) - \frac{shy}{sh(x + by)} \exp(-x - y) u_0(\xi) u_0(\eta), \tag{3.193}$$

where $u = u_0(\xi) u_0(\eta) R_{\infty 0}^{-1}(\xi, \eta, \varphi)$ and we have accounted for Eq. (3.86). We see that the problem of finding $R(\xi, \eta, \varphi, \tau)$ is reduced to the calculation of the previously studied functions $R_{\infty 0}(\xi, \eta, \varphi)$ and $u_0(\xi)$.

3.5.4 The Optically Thick Snow Layers with Arbitrary Level of Absorption

The equations for the arbitrary level of absorption can be presented in the following form (Sobolev 1975, 1984):

$$T(\xi, \eta, \tau) = \frac{M \exp(-k\tau)}{1 - N^2 \exp(-2k\tau)} u(\xi)u(\eta), \tag{3.194}$$

$$R(\xi, \eta, \varphi, \tau) = R_\infty - T(\xi, \eta)N \exp(-x), \tag{3.195}$$

where

$$M = 2 \int_{-1}^{1} i^2(\eta)\eta d\eta, \tag{3.196}$$

$$N = 2 \int_{0}^{1} u(\eta)i(-\eta)\eta d\eta. \tag{3.197}$$

The escape function $u(\eta)$ at any level of light absorption in the medium has been studied in detail by Yanovitskij (1997). It can be derived if the intensities in deep layers $i(\eta)$ and also the azimuthally averaged reflection functions of a semi-infinite turbid layer are known.

It follows that the simple relationships given above (at $b = 1$) can be derived from more general expressions given in this section under assumptions:

$$M = 1 - N^2, N = \exp(-y), u(\xi)u(\eta) = u_0(\xi)u_0(\eta), \tag{3.198}$$

which are valid as $\omega_0 \to 1$.

3.5.5 Account for Underlying Surface and Vertical Snow Inhomogeneity

Equations for finite layers given above can be applied for black underlying surface. An account for Lambertian surface with albedo A can be done in the following way (Liou 1992, 2002):

$$R(\xi, \eta, \tau) = R_{black}(\xi, \eta, \tau) + \frac{At(\xi)t(\eta)}{1 - Ar}. \tag{3.199}$$

Here r is the spherical albedo of a snow layer for the case of black underlying surface, $t(\xi)$ is total (direct + diffuse transmittance of a snow layer for the case of black underlying surface), $R_{black}(\xi, \eta, \tau)$ is the snow reflection function for the case of black underlying surface and we have assumed that a snow layer is vertically homogeneous. Therefore, $R(\xi, \eta, \tau)$ can be presented via known relationships valid for the case of black underlying surface. The derivation of Eq. (3.148) is simple. In particular, multiple reflections from the Lambertian surface can be presented in the following form (Stokes 1862):

$$R(\xi, \eta, \tau) = R_{black}(\xi, \eta, \tau) + t(\xi)A(1 + Ar + (Ar)^2 + \ldots)t(\eta). \qquad (3.200)$$

The summation of this series gives the result presented in Eq. (3.148). In a similar way one derives for the transmission function:

$$T(\xi, \eta, \tau) = T_{black}(\xi, \eta, \tau) + \frac{Ar_p(\xi)t(\eta)}{1 - Ar}. \qquad (3.201)$$

Snow is usually composed of sub-layers with various concentration of impurities and snow properties such as density, ice grain shapes and sizes. In the absence of absorption, the vertical inhomogeneity can be treated using the average values of the snow phase function/asymmetry parameter. However, if absorption must be taken into account, one must find the solution of the radiative transfer equation with vertically varying phase function, single scattering albedo, and extinction coefficient. The task must be solved using the numerical calculations.

Alternatively, the vertically inhomogeneous medium can be substituted by a homogeneous one with a special choice of effective single scattering albedo (Yanovit-skii 1991). The problem of two-layered snow with upper optically thick layer and lower semi-infinite layer with albedo r_∞ can be considered using equations given above. In particular, it follows in this case:

$$R(\xi, \eta, \tau) = R_u(\xi, \eta, \tau) + \frac{r_\infty t_u(\xi)t_u(\eta)}{1 - r_\infty r_u}, \qquad (3.202)$$

where index u signifies the upper layer. The account for the light scattering and absorption effects in atmosphere overlying the snowpack under assumption of the Lambertian reflective ground can be done with Eqs. (3.199), (3.201) as well. In this specific case the first terms in Eqs. (3.199), (3.201) must be substituted by the atmospheric reflection and transmission functions, respectively, and A has the meaning of the snow surface albedo. The plane albedo r_p, total transmittance t, and spherical albedo r must be changed to these parameters for the overlying atmosphere (Kokhanovsky et al. 2005b).

The correct treatment of the vertically inhomogeneous snow layers can be achieved by the numerical solution of the RTE for given vertical distribution of local optical characteristics of snow layers such as single scattering albedo, phase function, and extinction coefficient.

3.6 The Polarization of Light Reflected from Snow

The light reflected from the snow surface is polarized. It is characterized not only by its intensity I but also by other characteristics such as the degree of polarization P, the ellipticity ϵ, and the orientation of the polarization plane χ. These characteristics can be derived from the Stokes-vector—parameter $\vec{I}$ with the components I, Q, U, V:

$$P = \frac{\sqrt{Q^2 + U^2 + V^2}}{I}, \ \psi = \frac{1}{2}\arctan\left(\frac{U}{Q}\right), \ \varepsilon = \frac{1}{2}\arctan\left(\frac{V}{I}\right). \quad (3.203)$$

The Stokes—vector components can be expressed via the components of the electric vector $\vec{E}$ perpendicular (E_r) and parallel (E_l) to a selected plane (Stokes 1852):

$$I = E_l^* E_l + E_r^* E_r, \ Q = E_l^* E_l - E_r^* E_r, \ U = E_l^* E_r + E_r^* E_l,$$
$$V = i(E_r^* E_l - E_l^* E_r). \quad (3.204)$$

The ellipticity of reflected light is small and can be neglected for clean snow layers. Although the ellipticity of reflected light can be different from zero due to the presence of living cells in snow.

For most of applications, one can assume that light reflected from the snow is partially linearly polarized. The degree of polarization is small in the visible. However, it increases in the ice absorption bands. The example of the measured degree of linear polarization of light reflected from a semi-infinite snow layer is given in Fig. 3.15 for the solar zenith angle in the range 55–58 degrees. The illumination from the Sun comes from the backward direction (see Fig. 3.15) along the principal plane (PP), which contains the perpendicular to the surface and the direction towards the Sun (Tanikawa et al. 2014).

The degree of linear polarization is defined as

$$P = \frac{\Upsilon}{I}, \quad (3.205)$$

where $\Upsilon = \sqrt{Q^2 + U^2}$ is the polarized reflectance. One can also introduce the bidirectional polarized reflectance factor (BPRF) $\Pi = \pi \Upsilon / \xi F_0$, where πF_0 is the incident solar flux at the unit area perpendicular to the light beam and ξ is the cosine of the solar zenith angle.

The parameter U is close to zero in the principal plane (PP) containing the surface normal and the Sun direction (Peltoniemi et al. 2009). Therefore, the degree of linear polarization shown in Fig. 3.15 is derived from the measurements as follows:

$$P = -\frac{Q}{I}. \quad (3.206)$$

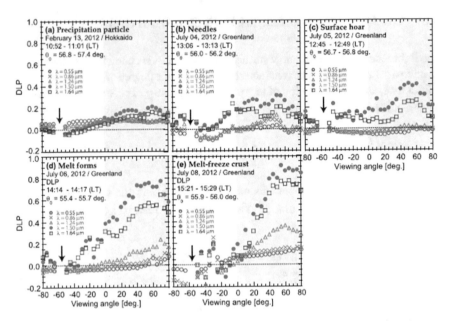

Fig. 3.15 Viewing angle dependence of the degree of linear polarization at five selected wavelengths. Positive viewing angles correspond to the forward direction and the negative ones correspond to the backward direction. The arrow on each figure shows the position of the Sun. The illumination from the Sun comes from the backward direction along the principal plane, which contains the perpendicular to the surface and the direction towards the Sun (Tanikawa et al. 2014)

The positive values of P correspond to the light polarization in the plane perpendicular to the principal plane ($\psi = \pi/2$). The negative values correspond to the polarization in the plane parallel to the scattering plane ($\psi = 0$). It follows from experimental data presented in Fig. 3.15 that light reflected from snow is weakly polarized in the visible. The degree of polarization increases in the near infrared, where light is polarized in the plane perpendicular to the principal plane for most of geometries. Although there are angular regions, where light is polarized in the plane parallel to the principal plane. It follows from Fig. 3.15 that there are neutral points, where light reflected form snow is completely unpolarized ($P = 0$). The lower degree of polarization in the visible as compared to the near infrared can be explained by the fact that the absorption of ice is small in the visible and multiple scattering dominates. Also light polarization by irregularly shaped ice crystals is weak (see Fig. 2.10). Multiple scattering leads to the increase of entropy and decrease of degree of polarization. Therefore, we conclude that spectral regions, where snow appear brighter are characterized by lower levels of degree of polarization. The peak at the viewing zenith angle around 55 degree (glint region) is due to the fact that the solar zenith angle is approximately equal to the Brewster angle $\varphi = \arctan(n)$, where n is ice refractive index, for the experiment presented. The melt-freeze crust is characterized by the largest values of the degree of polarization. It is a thin, glittering ice layer, which is sometimes visible on the snow on sunny days. Therefore, the polarimetric measurements at the Brewster geometry can be used to classify the snow conditions using remote sensing ground and satellite instrumentation.

The experimental measurements of light polarization by snow layers can be understood in terms of vector radiative transfer theory. In particular, it is known that the Stokes vector of light reflected and transmitted by snow layers is satisfied to the following vector radiative transfer equation (VRTE) (Liou 1992):

$$\frac{d\vec{I}(\vec{\Omega})}{dl} = -\hat{\sigma}_{ext}\,\vec{I}(\vec{\Omega}) + \int_{4\pi} \hat{L}_2\hat{\sigma}_{sca}\left(\vec{\Omega}, \vec{\Omega}'\right)\hat{L}_1\,\vec{I}(\vec{\Omega}')d\,\vec{\Omega}'. \qquad (3.207)$$

The matrix $\hat{L}_1$ transforms the Stokes vector defined in the meridional plane holding the normal to the scattering layer and the direction $\vec{\Omega}'$ to the Stokes vector in the scattering plane. The matrix $\hat{L}_2$ is needed for the transformation of the Stokes vector of scattered light from the scattering plane back to the meridional plane (see Appendix 2). This is due to the fact that the matrix $\hat{\sigma}_{sca}$ in Eq. (3.194) is defined in the scattering plane and the vector $\vec{I}$ is defined with respect to the meridional plane. For randomly oriented ice grains the matrix $\hat{\sigma}_{ext}$ is reduced to a scalar value k_{ext} and

$$\frac{d\vec{I}(\vec{\Omega})}{dl} = -k_{ext}\,\vec{I}(\vec{\Omega}) + \int_{4\pi} \hat{L}_2\hat{\sigma}_{sca}\left(\vec{\Omega}, \vec{\Omega}'\right)\hat{L}_1\,\vec{I}(\vec{\Omega}')d\,\vec{\Omega}'. \qquad (3.208)$$

We will be mostly concerned with solutions of Eq. (3.195) for a plane-parallel snow layer illuminated by the Sun. The interaction of solar radiation with extended snow fields is well-approximated by the solution of this idealized problem. The solar light with the zenith angle ϑ_0 uniformly illuminates a plane-parallel scattering layer from above. We will assume that properties of the layer do not change in the horizontal direction. Then properties of light field depend only on the vertical coordinate Z and the direction $\vec{\Omega}$, specified by the zenith angle ϑ and the azimuth φ. The main task is to calculate angular distributions $\vec{I}(\vartheta, \varphi, z)$. Usually only measurements of $\vec{I}(\vartheta, \varphi, 0)$ at the top of the snow (reflected light) are performed. Therefore, we will be concerned with the angular distribution of the reflected light.

Equation (3.195) takes the following form for the plane-parallel snow layer:

$$\cos\vartheta\,\frac{d\vec{I}(\vartheta, \varphi)}{dz} = -\sigma_{ext}\,\vec{I}(\vartheta, \varphi)$$

$$+ \int_0^{2\pi} d\varphi' \int_0^{\pi} d\vartheta'\hat{L}_2\hat{\sigma}_{sca}\left(\vartheta', \varphi' \to \vartheta, \varphi\right)\hat{L}_1\,\vec{I}(\vartheta', \varphi'). \qquad (3.209)$$

The most simple case to study is that of idealized homogeneous snow layers with values of k_{ext} and $\hat{\sigma}_{sca}$ not dependent on the position inside the cloud. Then we have from Eq. (3.198):

$$\cos\vartheta\,\frac{d\overrightarrow{I}\,(\vartheta,\varphi)}{d\tau} = -\overrightarrow{I}\,(\vartheta,\varphi) + \frac{\omega_0}{4\pi}\int\limits_0^{2\pi} d\varphi'\int\limits_0^{\pi} d\vartheta'\hat{L}_2\hat{P}\big(\vartheta',\varphi'\to\vartheta,\varphi\big)\hat{L}_1\overrightarrow{I}(\vartheta',\varphi'),$$

$$(3.210)$$

where $\tau = k_{ext}z$ is the optical depth,

$$\hat{P} = \frac{4\pi\hat{\sigma}_{sca}\big(\vartheta',\varphi'\to\vartheta,\varphi\big)}{k_{sca}} \qquad (3.211)$$

is the phase matrix and

$$\omega_0 = \frac{k_{sca}}{k_{ext}} \qquad (3.212)$$

is the single scattering albedo. The phase matrix, extinction coefficient and single scattering albedo of snow grains have been studied in Chap. 2.

It is useful to distinguish between diffused $\overrightarrow{I}_d$ and direct (or coherent) $\overrightarrow{I}_c = \overrightarrow{F}_0\,\delta(\mu-\mu_0)\delta(\varphi-\varphi_0)$ light in the general solution $\overrightarrow{I}\,(\vartheta,\varphi)$. Here $\overrightarrow{F}_0$ describes the Stokes vector of the incident attenuated flux. It is assumed that the layer is illuminated in the direction defined by the incidence zenith angle $\vartheta_0 = \arccos(\mu_0)$ and the azimuthal angle φ_0. The multiply scattered light is observed in the direction specified by the zenith observation angle $\vartheta = \arccos(\mu)$ and the azimuth φ. Namely, we write:

$$\overrightarrow{I}\,(\vartheta,\varphi) = \overrightarrow{I}_d(\vartheta,\varphi) + \overrightarrow{I}_c(\vartheta,\varphi). \qquad (3.213)$$

The substitution of this formula in Eq. (3.199) gives

$$\cos\vartheta\,\frac{d\overrightarrow{I}_d(\vartheta,\varphi)}{d\tau} = -\overrightarrow{I}_d(\vartheta,\varphi)$$

$$+ \frac{\omega_0}{4\pi}\int\limits_0^{2\pi} d\varphi'\int\limits_0^{\pi} d\vartheta'\hat{L}_2\hat{P}\big(\vartheta',\varphi'\to\vartheta,\varphi\big)\hat{L}_1\overrightarrow{I}_d(\vartheta',\varphi')$$

$$+ \frac{\omega_0}{4}\hat{P}_*(\vartheta_0,\varphi_0\to\vartheta,\varphi)\overrightarrow{F}_0\exp\left(-\frac{\tau}{\cos\vartheta_0}\right). \qquad (3.214)$$

Here $\hat{P}_*$ is the correspondent scattering matrix defined with respect to the meridional plane, $\overrightarrow{F}$ (F_0, Q_0, U_0, V_0) is the Stokes vector of incident light flux at the top of a cloud. It follows for unpolarized solar light: $Q_0 = U_0 = V_0$. The solution of this equation under boundary conditions stating that there is no diffused light entering the cloud from above and below, allows to find $\overrightarrow{I}_d(\vartheta,\varphi)$. Also it follows that $\overrightarrow{I}_c(\vartheta,\varphi)$

is given simply by:

$$\vec{I}_c(\vartheta, \varphi) = \vec{F}(\vartheta, \varphi)\delta(\cos\vartheta - \cos\vartheta_0)\delta(\varphi - \varphi_0)\exp\left(-\frac{\tau}{\cos\vartheta_0}\right). \quad (3.215)$$

The solution of Eq. (3.213) is a more easy task than that of Eq. (3.199) because we avoid the necessity to deal with the divergence in the direction of incident light existing in the framework of Eq. (3.199). There are multiple numerical algorithms, which can be used to solve the integro—differential VRTE (3.213) accounting also for the vertical variation of the snow properties (http://en.wikipedia.org/wiki/Atmospheric_radiative_transfer_codes). The input for the corresponding codes in the case of vertically homogeneous snow layers is the optical thickness, single scattering albedo and the phase matrix of snow grains, which depend on the size of grains, their shapes and wavelength (Siewert 2000).

Equations for optically thick layers presented in the previous section can be generalized to account for polarization. Correspondent equations for azimuthally averaged reflection $\hat{R}$ and transmission $\hat{T}$ matrices were obtained by Domke (1978a, b). They have the following forms for isotropic symmetric light scattering media:

$$\hat{R}(\xi, \eta) = \hat{R}_\infty(\xi, \eta) - N\hat{T}(\xi, \eta)\exp(-k\tau), \quad (3.216)$$

$$\hat{T}(\xi, \eta) = \frac{M\exp(-k\tau)}{1 - N^2\exp(-2k\tau)}\vec{K}(\eta)\vec{K}^T(\xi), \quad (3.217)$$

where only two-dimensional matrices and vectors are involved. Other components of generally four-dimensional matrices and vectors vanish due the azimuthal averaging. Note that this is also the case for a normal illumination of a snow layer. Then the azimuth does not enter theory at all. Here $\hat{R}_\infty(\xi, \eta)$ is the azimuthally averaged reflection matrix of a semi-infinite medium with the same optical characteristics as a finite slab under study. One can also introduce the vector $\vec{\Theta}(\mu)$, which describes the intensity and degree of light polarization in deep layers of a semi-infinite scattering medium [in a so-called asymptotic regime, when the intensity and polarization angular distributions are symmetrical with respect to the normal to a scattering layer and exponentially decrease with the depth ($\sim \exp(-k\tau)$)]. Functions $\hat{R}_\infty(\xi, \eta)$ and $\vec{\Theta}(\eta)$ determine all parameters in the equations given above. The explicit equations for the calculation of $\hat{R}_\infty(\xi, \eta)$ and $\vec{\Theta}(\eta)$ and also the escape vector $\vec{K}(\eta)$ are discussed by Domke (1978a, b, 2016).

We see, therefore, that intensity and polarization characteristics of reflected and transmitted light for optically thick snow layers are determined by the reflection matrix of a semi-infinite layer and the angular distribution of the light intensity and polarization in deep layers of the same medium. This reduction of a problem for a finite optical thick snow layer to the case of a semi-infinite medium is of a general importance for the radiative transfer theory. Note that the matrix $\hat{R}_\infty$ and vector $\vec{\Theta}$

are obtained from solutions of the well-known integral equations, which can be found elsewhere (van de Hulst 1980; de Rooij 1985).

Equations (3.215), (3.216) are valid only for the azimuthally averaged matrices. In practice, however, measurements are performed for a fixed azimuth. The transmission matrix is azimuthally independent in the case of optically thick layers. The azimuthal dependence in the reflected light disappears in some specific cases (e.g., for the case of normal illumination of a flat snow surface).

Equations (3.215), (3.216) are simplified for nonabsorbing media. Then it follows:

$$\hat{R}(\xi, \eta) = \hat{R}_v(\xi, \eta) - \hat{T}(\xi, \eta), \tag{3.218}$$

$$\hat{T}(\xi, \eta) = \frac{4}{3(\tau + 2q_0)(1 - g)} \vec{K}_0(\xi) \vec{K}_0^T(\eta), \tag{3.219}$$

where

$$q_0 = \frac{2}{1 - g} \int_0^1 d\eta \eta^2 \vec{K}_0^T(\eta) \vec{j}. \tag{3.220}$$

Here

$$\vec{j} = \begin{pmatrix} 1 \\ 0 \end{pmatrix} \tag{3.221}$$

is the unity vector, g is the asymmetry parameter,

$$\vec{K}_0(\eta) = \frac{3}{4} \left[\eta + 2 \int_0^1 d\xi \xi^2 \hat{R}_\infty^0(\xi, \eta) \right] \vec{j} \tag{3.222}$$

and $\hat{R}_\infty^0(\mu, \mu_0)$ is the azimuthally averaged reflection matrix of a semi-infinite nonabsorbing medium. This matrix is completely determined by the phase matrix $\hat{P}$, introduced above. It does not depend on the single scattering albedo and optical thickness by definition. Clearly, the first component of the vector $\vec{K}_0$ coincides with the escape function u_0 discussed above.

These asymptotic equations are simple in form. However, they can be used only if auxiliary functions and parameters are known. Their calculations, however, can be quite a complex procedure. However, it appears that for weakly absorbing media, when single scattering albedo is close to one, simplifications are possible. Then it follows (Kokhanovsky 2003):

$$\hat{R}(\xi, \eta) = \hat{R}_{0\infty}(\xi, \eta) \exp\left(-y\hat{D}(\xi, \eta)\right) - \hat{T}(\xi, \eta) \exp(-x - y), \tag{3.223}$$

$$\hat{T}(\xi, \eta) = t \, \overrightarrow{K}_0(\xi) \, \overrightarrow{K}_0^T(\eta), \qquad (3.224)$$

where $x = k\tau$, $y = 4\sqrt{\frac{1-\omega_0}{3(1-g)}}$, $k = \sqrt{3(1-\omega_0)(1-g)}$, $\hat{D}(\xi, \eta) =$
$\hat{R}_{0\infty}^{-1}(\xi, \eta) \, \overrightarrow{K}_0(\xi) \, \overrightarrow{K}_0^T(\eta)$, $t = \frac{\sinh y}{\sinh(x+\alpha y)}$ is the global transmittance of a scattering
layer, $\alpha = \frac{1}{2} \int_0^1 u_0(\eta)\eta^2 d\eta \approx 1.07$, and $\hat{R}_{0\infty}(\mu, \mu_0)$ is the reflection matrix of a semi-
infinite nonabsorbing layer with the same phase matrix as an absorbing layer of a
finite thickness under study. The two-dimensional vector $\overrightarrow{K}_0(\mu)$ describes the polarization and intensity of light in the Milne problem for nonabsorbing semi-infinite
media (Wauben 1992). The components $K_{01}(\mu)$ and $K_{02}(\mu)$ of this vector were
calculated by Chandrasekhar (1950) for Rayeigh particles ($g = 0$) and by Wauben
(1992) for spherical particles with the refractive index $n = 1.44$ and the gamma
particle size distribution (1.5) with $\mu = 11.3$, $a_0 = 0.83$ μm. The wavelength of λ
was equal to 0.55 μm. Note, that the model of spheres with $\mu = 11.3$, $r_0 = 0.83$ μm,
and $n = 1.44$ is generally used to characterize particles in clouds on Venus (Hansen
and Travis 1974). It follows for the effective size a_{ef}, the effective variance Δ_{ef}, and
the asymmetry parameter g, respectively, in this case: $\Delta_{ef} = 1.05\mu m$, $\Delta_{ef} = 0.07$,
$g = 0.718$. It was found that the ratio $p_l = -\frac{K_{02}}{K_{01}}$, which gives the degree of polarization for transmitted light is very low. It changes from zero to 1.2% while the
escape angle changes from 0 till 90 degrees. Note that for Rayleigh scattering we
have a change from 0 till 11.7% for the same conditions. This means that light transmitted by thick snow layers is almost unpolarized. It is possible to understand this
on general grounds. Indeed, the polarization of unpolarized solar light occurs due to
single scattering events. Multiple light scattering leads to an increase of entropy and
the reduction of initial polarization arising in single scattering events.

Equations given above can be simplified for nonabsorbing media ($y = 0$):

$$\hat{R}(\xi, \eta) = \hat{R}_{0\infty}(\xi, \eta) - \hat{T}(\xi, \eta), \qquad (3.225)$$

$$\hat{T}(\xi, \eta) = t \, \overrightarrow{K}_0(\xi) \, \overrightarrow{K}_0^T(\eta), \qquad (3.226)$$

where

$$t = \frac{1}{\alpha + \frac{3}{4}\tau(1-g)} \qquad (3.227)$$

is the global transmittance.

Let us apply Eq. (3.225) to a particular problem, namely, to the derivation of a
relation between the spherical albedo $r = 1 - t$ and the degree of polarization of
reflected light $p_l(\mu)$ at the illumination along the normal to the snow layer ($\mu_0 = 1$)

by a wide, unidirectional unpolarized light beam. The value of $p_l(\mu)$ is given simply by $-R_{21}(1, \mu)/R_{11}(1, \mu)$ in this case. Thus, it follows from Eq. (3.225):

$$p_l(\xi) = \frac{p_{l\infty}(\xi)}{1 - (1 - r)\mathbb{N}(\xi)},$$ (3.228)

where

$$\mathbb{N}(\xi) = \frac{u_0(1)u_{0\xi}(\mu)}{R_{0\infty}(1, \mu)},$$ (3.229)

$$p_{l\infty}(\xi) = -\frac{R_{\infty 21}(1, \xi)}{R_\infty(1, \xi)}$$ (3.230)

and we accounted for the equality: $K_{02}(1) = 0$. It appears that the value of $\mathbb{N}(\xi)$ is close to 1 for most of observation angles, which implies the inverse proportionality between the brightness of a turbid medium and the degree of polarization of reflected light ($rp_l \approx p_{l\infty}$). This inverse proportionality between the spherical albedo r and the degree of polarization p_l was discovered experimentally by Umow (1905). Equation (3.228) can be considered as a manifestation of this important law, which has important applications in reflectance spectroscopy (Hapke 1993).

Equation (3.228) is easily generalized to account for the absorption of light in a medium using the exponential approximation described above. Namely, it follows:

$$p_l(\xi) = \frac{p_{l\infty}^*(\xi)}{1 - \mathbb{N}^*(\xi)t \exp(-x - y)},$$ (3.231)

where

$$t = \frac{\sinh y}{\sinh(x + \sigma y)}$$ (3.232)

and

$$\mathbb{N}^*(\xi) = \frac{u_0(1)u_0(\xi)}{R_\infty^*(1, \xi)}.$$ (3.233)

Values of $p_{l\infty}^*(\xi)$ and $R_\infty^*(1, \xi)$ represent the degree of polarization and reflection function of a semi-infinite weakly absorbing medium at the nadir illumination. Note, that Eq. (3.317) can be written in the following form:

$$p_l(\xi) = c(\xi, \tau)p_{l\infty}^*(\xi),$$ (3.234)

where

$$c(\xi, \tau) = \frac{1}{1 - \mathbb{N}^*(\xi) t \exp(-x - y)} \tag{3.235}$$

can be interpreted as the polarization enhancement factor, which is solely due to a finite snow depth. It follows for semi-infinite layers that the transmittance t is equal to zero and $c = 1$ as it should be. Also it follows from Eq. (3.320) that zeroes of polarization curves for semi-infinite and optically thick finite snow layers almost coincide. This is due to the fact that the function $\mathbb{N}^*(\xi)$ only weakly depends on the angle. Multiple light scattering fails to produce the polarization of incident unpolarized light. It only diminishes the polarization of singly scattered light. Thus, the angles where polarization is equal to zero for semi-infinite layers are almost equal to those for the case of single light scattering.

In the case of semi-infinite weakly absorbing snow layers, it follows (Kokhanovsky 2003):

$$\hat{R}_\infty(\xi, \eta) = \hat{R}_{0\infty}(\xi, \eta) \exp\left(-y \hat{D}(\xi, \eta)\right) \tag{3.236}$$

or assuming that $y \to 0$:

$$\hat{R}(\xi, \eta) = \hat{R}_{0\infty}(\xi, \eta) - y \, \vec{K}_0(\xi) \, \vec{K}_0^T(\eta). \tag{3.237}$$

Let us write Eq. (3.237) in the explicit form:

$$\begin{pmatrix} R_{\infty 11} & R_{\infty 12} \\ R_{\infty 21} & R_{\infty 22} \end{pmatrix} = \begin{pmatrix} R_{0\infty 11} & R_{0\infty 12} \\ R_{0\infty 21} & R_{0\infty 22} \end{pmatrix} - y \begin{pmatrix} K_{01}(\xi) K_{01}(\eta) & K_{01}(\xi) K_{02}(\eta) \\ K_{02}(\xi) K_{01}(\eta) & K_{02}(\xi) K_{02}(\eta) \end{pmatrix}. \tag{3.238}$$

In the case of unpolarized solar light illumination we need to multiple Eq. (3.238) by the column vector (van de Hulst 1980)

$$\vec{j} = \begin{pmatrix} 1 \\ 0 \end{pmatrix}. \tag{3.239}$$

Then it follows:

$$\begin{pmatrix} R_{11} \\ R_{21} \end{pmatrix} = \begin{pmatrix} R_{\infty 11} \\ R_{\infty 21} \end{pmatrix} - y \begin{pmatrix} K_{01}(\xi) K_{01}(\eta) \\ K_{02}(\xi) K_{01}(\eta) \end{pmatrix} \tag{3.240}$$

or

$$R_{\infty 11} = R_{0\infty 11} - y K_{01}(\xi) K_{01}(\eta), \tag{3.241}$$

$$R_{\infty 21} = R_{\infty 021} - y K_{02}(\xi) K_{01}(\eta). \tag{3.242}$$

Equation (2.241) gives the relationship between the reflection function of absorbing ($R_\infty = R_{\infty 11}$) and nonabsorbing ($R_{0\infty} = R_{0\infty 11}$) snow layers as $y \to 0$. The second equation shows the difference of reflection functions for the polarizations parallel (R_l) and perpendicular (R_r) to the principal plane:

$$R_{\infty 21} = R_{\infty l} - R_{\infty r} = R_{\infty 021} - y K_{02}(\xi) K_{01}(\eta). \tag{3.243}$$

Also it follows for the degree of polarization:

$$p_\infty = p_{0\infty} + \Lambda y, \tag{3.244}$$

where

$$\Lambda = \frac{K_1(\xi) K_1(\eta)}{R_{0\infty}} p_{0\infty} + \frac{K_2(\xi) K_1(\eta)}{R_{0\infty}}. \tag{3.245}$$

Taking into account that $K_2(\xi)$ describes the degree of polarization in deep layers of nonabsorbing media, we can assume that $K_2(\xi) \ll 1$ and derive from Eqs. (2.244), (2.245):

$$p_\infty = p_{0\infty}\left(1 + y\frac{K_1(\xi) K_1(\eta)}{R_{0\infty}}\right) \tag{3.246}$$

or

$$p_\infty = p_{0\infty} \exp\left[y\frac{K_1(\xi) K_1(\eta)}{R_{0\infty}}\right]. \tag{3.247}$$

It follows from Eq. (2.247) that the degree of polarization increases with the level of light absorption in the medium. This is confirmed by the experimental data shown in Fig. 3.15. It should be pointed out that the approximation of spherical particles can not be used in snow polarimetry. The peak in the degree of polarization due to enhanced light polarization in the rainbow region common for large spherical particles does not exist for snow surfaces. The dependence of the degree of polarization on the viewing zenith angle calculated using Eq. (2.247) is presented in Fig. 3.16. The results presented in Fig. 3.16 closely reproduce experimentally derived features of DOLP given in Fig. 3.15 such as low values of degree of polarization for snow, the decrease of degree of polarization towards the antisolar point (-57 degrees) and neutral point around -40 degrees for the SZA = 57 degrees.

The experimentally measured and calculated using Eq. (2.247) spectral degree of polarization is shown in Fig. 3.17. The value of $p_{0\infty}$ has been calculated using single scattering approximation. For the nadir illumination case as shown in Fig. 3.17 the

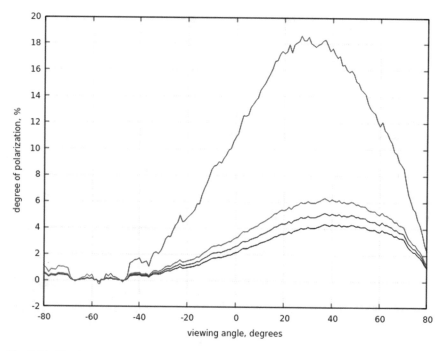

Fig. 3.16 The degree of polarization calculated using Eq. (2.247) at the SZA $= 57$ degrees and the fractal model of ice grains (Macke et al. 1996) under assumption that $y = \sqrt{\alpha L}$, where α is the bulk ice absorption coefficient and $L = 1$ mm. The wavelengths are 0.55 (lower curve), 1.02, 1.24, and 1.6 μm (upper curve) with larger wavelengths corresponding to the larger values of degree of polarization. The calculations are performed in the principal plane with negative viewing angles corresponding to the backscattering direction

value of $p_{0\infty}$ coincides with the normalized phase matrix element P_{12} (see Appendix 2) for nonabsorbing ice crystals in the approximation under study. We have used the geometrical optics calculations of P_{12} performed by Macke et al. (1996) for fractal ice grains.

It follows that Eq. (2.247) provides means for the calculation of the snow degree of polarization at channels, where the snow can be considered as a weakly absorbing medium (spherical albedo above 0.1). The theoretical results around 2000 nm are not reliable due to the fact that snow can be not be considered as a weakly absorbing medium at this channel. Taking into account that it follows for the reflection function:

$$R_{\infty} = R_{0\infty} \exp\left[-y \frac{K_1(\xi) K_1(\eta)}{R_{0\infty}}\right], \tag{3.248}$$

we conclude that the second Stokes parameter

$$Q = -p_{\infty} R_{\infty} = -p_{0\infty} R_{0\infty} \tag{3.249}$$

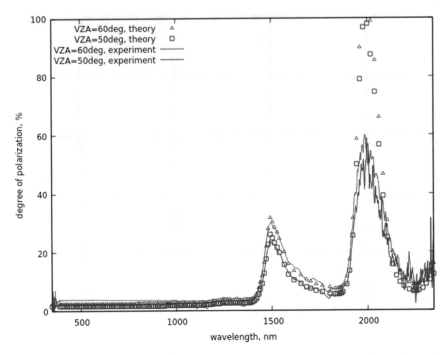

Fig. 3.17 The spectral degree of polarization calculated using Eq. (2.247) at the VZA $= 50$ and 60 degrees at the nadir illumination in the principal plane (forward scattering) and the fractal model of ice grains (Macke et al. 1996) under assumption that $y = \sqrt{\alpha L}$, where α is the bulk ice absorption coefficient and $L = 1mm$. The results of experimental measurements (Sun et al. 2021) are also shown

does not depend on the single scattering albedo for the wavelengths, where snow is weakly absorbing. This is also confirmed by experimental data (Sun et al. 2021).

It follows from Eq. (3.249):

$$p_\infty = \frac{A}{R_\infty}, \tag{3.250}$$

where $A = p_{0\infty} R_{0\infty}$. Equation (3.250) is the manifestation of the Umow law, which state that the degree of polarization is inversely proportional to the brightness of the turbid medium.

References

Ambartsumian, V.A. 1943. On the diffuse reflection of light from turbid media. *Doklady AN SSSR* 8: 257–261.

Ambartsumian, V.A. 1944. On the one-dimensional case of the problem related to the light scattering and absorbing medium of the finite optical thickness. *Izvestiya, Academy of Sciences of Armenian SSR, Natural Sciences* 1–2: 31–36.

Ambartsumian, V.A. 2011. Selected papers. *Stars, nebulae, and galaxies*, Advances in Astronomy and Astrophysics, vol.10, ed. G. Meylan. Cambridge Scientific Publishers.

Aoki, T., T. Aoki, M. Fukabori, and A. Uchiyama. 1999. Numerical simulation of the atmospheric effects on snow albedo with a multiple scattering radiative transfer model for the atmosphere-snow system. *Journal of the Meteorological Society of Japan* 77: 595–614.

Aoki, T., Kuchiki, K., Niwano, M., Kodama, Y., Hosaka, M., and Tanaka, T. 2011. Physically based snow albedo model for calculating broadband albedos and the solar heating profile in snowpack for general circulation models. *Journal Geophysical Research* 116:, D11114. https://doi.org/10.1029/2010JD015507.

Chandrasekhar, S. 1943. Stochastic problems in physics and astronomy. *Reviews of Modern Physics* 1: 1–89.

Chandrasekhar, S. 1950. *Radiative transfer.* Oxford: Oxford Press.

de Rooij, W.A. 1985. *Reflection and transmission of polarized light by planetary atmospheres.* PhD. thesis, Free University of Amsterdam.

Deirmendjian, A. 1969. *Electromagnetic scattering on spherical polydispersions.* Amsterdam: Elsevier.

Domke, H. 1978a. Linear fREDHOLM integral equations for radiative transfer problems in finite plane-parallel media. I. Imbedding in an infinite medium. *Astronomische Nachrichten* 299: 87–93.

Domke, H. 1978b. Linear fREDHOLM integral equations for radiative transfer problems in finite plane-parallel media. II. Imbedding in a semi- infinite medium. *Astronomische Nachrichten* 299: 95–102.

Domke, H. 2016. On inhomogeneous atmospheres—on transforming conservative multiple scattering to non-conservative multiple pseudo—scattering. *Journal of Quantitative Spectroscopy & Radiative Transfer* 183: 56–63.

Feynman, R., and A. R. Hibbs. 1965. *Quantum mechanics and path integrals.* N.Y.: McGraw-Hill.

Hale, G.M., and M.R. Querry. 1973. Optical constacts of water in the 200-nm to 200-μm wavelength region. *Applied Optics* 12: 555–563.

Hansen, J.E., and L.D. Travis. 1974. Light scattering in planetary atmospheres. *Space Science Reviews* 16: 527–610. https://doi.org/10.1007/BF00168069.

Hapke, B. 1993. *Theory of the reflectance and emittance spectroscopy.* Cambridge: University Press.

Kokhanovsky, A.A., and E.P. Zege. 2004. Scattering optics of snow. *Applied Optics* 43: 1589–1602.

Kokhanovsky, A.A., M. Lamare, B. Di Mauro, G. Picard, L. Arnaud, M. Dumont, F. Tuzet, C. Brockmann, and J.E. Box. 2018. On the reflectance spectroscopy of snow. *The Cryosphere* 12: 2371–2382.

Kokhanovsky, A.A., T. Aoki, A. Hachikubo, M. Hori, and E.P. Zege. 2005a. Reflective properties of natural snow: Approximate asymptotic theory versus in situ measurements. *IEEE Transactions on Geoscience and Remote Sensing* 43: 1529–1535.

Kokhanovsky, A., B. Mayer, and V.V. Rozanov. 2005b. A parameterization of the diffuse transmittance and reflectance for aerosol remote sensing problems. *Atmospheric Research* 73: 37–43.

Kokhanovsky, A.A. 2003. *Polarization optics of random media.* Berlin: Springer-Praxis.

Kokhanovsky, A.A. 2002. Statistical properties of photon gas in random media. *Physical Review* E66, 037601.

Kokhanovsky, A.A. 2005. Reflection of light from particulate media with irregularly shaped particles. *Journal of Quantitative Spectroscopy and Radiation Transfer* 96: 1–10.

Kokhanovsky, A., B. Mayer, W. von Hoyningen-Huene, S. Schmidt, and P. Pilewskie. 2007. Retrieval of cloud spherical albedo from top-of-atmosphere reflectance measurements performed at a single observation angle. *Atmospheric Chemistry and Physics* 7: 3633–3637.

Kokhanovsky, A.A., M. Lamare, O. Danne, et al. 2019. Retrieval of snow properties from the Sentinel-3 ocean and land colour instrument. *Remote Sensing* 11: 2280. https://doi.org/10.3390/rs11192280.

Kokhanovsky, A.A., J.E. Box, B. Vandecrux, K.D. Mankoff, M. Lamare, A. Smirnov, and M. Kern. 2020. The determination of snow albedo from satellite measurements using fast atmospheric correction technique. *Remote Sensing* 12: 234. https://doi.org/10.3390/rs12020234.

Liou, K.N. 1992. *Radiation and cloud processes in atmosphere.* Oxford: Oxford University Press.

Liou, K.N. 2002. *An introduction to atmospheric radiation.* N.Y.: Academic Press.

Macke, A., and F. Tzschihholz. 1992. Scattering of light by fractal particles: A qualitative estimate exemplary for two-dimensional triadic Koch island. *Physica A: Statistical Mechanics and Its Applications* 191: 159–170.

Macke, A., et al. 1996. Scattering properties of atmospheric ice crystals. *Journal of Atmospheric Science* 53: 2813–2825.

Minin, I.N. 1988. *Radiative transfer theory in planetary atmospheres.* Moscow: Nauka.

Mishchenko, M.I., J.M. Dlugach, E.G. Yanovitskij, and N.T. Zakharova. 1999. Bidirectional reflectance of flat, optically thick particulate layers: An efficient radiative transfer solution and applications to snow and soil surfaces. *Journal of Quantitative Spectroscopy & Radiative Transfer* 63: 409–432.

Peltoniemi, J., T. Hakala, J. Suomalainen, E. Puttonen. 2009. Polarised bidirectional reflectance factor measurements from soil, stones, and snow. *Journal of Quantitative Spectroscopy and Radiative Transfer* 110(17): 1940–1953. ISSN 0022-4073. https://doi.org/10.1016/j.jqsrt.2009.04.008; https://www.sciencedirect.com/science/article/pii/S0022407309001538.

Perelman, L.T., J. Wu, I. Itzkan, and A.S. Feld. 1994. Photon migration in turbid media using path integrals. *Physical Review Letter* 72: 1341–1344.

Perovich D.K. 2007. Light reflection and transmission by a temperate snow cover. *Journal of Glaciology* 53(181): 201–210.

Picard, G., M. Dumont, M. Lamare, F. Tuzet, F. Larue, R. Pirazzini, and L. Arnaud. 2020. 2020: Spectral albedo measurements over snow-covered slopes: Theory and slope effect corrections. *The Cryosphere* 14: 1497–1517. https://doi.org/10.5194/tc-14-1497-2020.

Ricchiazzi, P., S. Yang, C. Gautier, C., and D. Sowle, 1998. SBDART: A research and teaching tool for plane-parellel radiative transfer in the Earth's atmosphere. *Bulletin of the American Meteorological Society* 79: 2101–2114

Rosenberg, G.V. 1962. Optical characteristics of thick weakly absorbing scattering layers. *Doklady Akademii Nauk* 6: 775–777.

Shettle, E.P., and R.W. Fenn. 1979. Models for the aerosols of the lower atmosphere and the effects of humidity variations on their properties. Report AFGt-TR-79-O21, Air Force Geophysical Laboratory, USA.

Siewert, C.E. 2000. A discrete-ordinates solution for radiative—transfer models that include polarization effects. *Journal of Quantitative Spectroscopy & Radiative Transfer* 64: 227–254.

Sobolev, V.V. 1975. *Light scattering in planetary atmospheres.* N.Y.: Pergamon Press.

Sobolev, V.V. 1984. Integral relations and asymptotic expressions in the theory of radiative transfer. *Astrofizika* 20: 123–132.

Stokes, G.G. 1852. On the composition and resolution of streams of polarized light from different sources. *Transactions of the Cambridge Philosophical Society* 9: 399–416.

Stokes, G.G. 1862. On the intensity of the light reflected from or transmitted through a pile of plates. *Proceedings of the Royal Society of London* 11: 545–556.

Sun, J., D. Wu, and Y. Lv. 2021. Optical properties of snow surfaces: multi-angular photometric and polarimetric hyperspectral measurements. *IEEE Transactions on Geoscience and Remote Sensing.* https://doi.org/10.1109/TGRS.2021.3078170.

Tanikawa, T., M. Hori, T. Aoki, A. Hachikubo, K. Kuchiki, M. Niwano, M., S. Matoba, S. Yamaguchi, and K. Stamnes. 2014. In situ measurements of polarization properties of snow surface under the Brewster geometry in Hokkaido, Japan, and northwest Greenland ice sheet. *Journal of Geophysical Research: Atmospheres* 119: 13,946–13,964. https://doi.org/10.1002/2014JD022325

Umow, N. 1905. Chromatische Depolarisation durch Lichtzerstreung. *Physikalishce Zeitschrift* 6: 674–676.

van de Hulst, H.C. 1980. *Multiple light scattering: Tables, formulas and applications.* N.Y.: Academic Press.

Warren, S.G. 1984. Optical constants of ice from the ultraviolet to the microwave. *Applied Optics* 23: 1206–1225.

Wauben, W.M.F. 1992: *Multiple Scattering of Polarized Radiation in Planetary Atmospheres, PhD thesis,* Free University of Amsterdam.

Yanovitskij, E.G. 1997. *Light scattering in inhomogeneous atmospheres.* N.Y.: Springer.

Zege, E.P., A.P.Ivanov, and I.L. Katsev. 1991. *Image transfer through a scattering medium.* N.Y.: Springer.

Zhuravleva, T.B., and A.A. Kokhanovsky. 2011. Influence of surface roughness on the reflective properties of snow. *Journal of Quantitative Spectroscopy and Radiative Transfer* 112 (8): 1353–1368. https://doi.org/10.1016/j.jqsrt.2011.01.004.

Chapter 4
Remote Sensing of Snow

4.1 Determination of Local Optical Parameters of Snow

4.1.1 Semi-Infinite Snow Layers

Spectroscopy is used to refer to the measurement of transmitted/reflected radiation intensity as a function of wavelength. The intensity of light transmitted/reflected by snow layers depends on local optical parameters of snow such as extinction coefficient, absorption coefficient, and directional light scattering coefficient (phase function). In turn these local optical parameters depend on the snow microstructure and pollution load. Therefore, several important snow parameters can be derived from snow spectral reflectance and transmittance measurements. In some cases the spectral polarization parameters are measured. Usually, the snow reflectance spectroscopy is used due to much easier experimental setup as compared to the snow transmittance/internal light filed measurements. However, diffuse light transmittance measurements enhance the information content and, therefore, provide additional information for the solution of inverse problems of snow optics. Therefore, we consider the techniques for the determination of snow properties based on both reflectance and transmittance measurements.

We shall start from the determination of snow local optical characteristics using spectral reflectance and transmittance measurements. Snow microstructure parameters and pollution load can be derived, if snow spectral optical characteristics are known. The main optical characteristics of snow are the extinction coefficient, absorption coefficient, and phase function.

Snow is composed of ice crystals. Therefore, the snow phase function is close to that of crystalline clouds. Phase functions of various crystalline clouds have been studied experimentally. They can be modelled by the combination of exponentials or alternatively by the combination of the Henyey–Greenstein phase functions as discussed in Chap. 2. Clearly, the morphology of snow covers is not identical to that of crystalline clouds due the snow metamorphism processes and external factors such as wind, ground temperature variations, solar insolation, and precipitation including

© Springer Nature Switzerland AG 2021
A. Kokhanovsky, *Snow Optics*,
https://doi.org/10.1007/978-3-030-86589-4_4

rain. However, one may expect that phase functions of snow are close to that if ice clouds with irregularly shaped crystals measured in the laboratory. The main feature of snow phase function is its smooth behavior in the backward region. The asymmetry parameter of ice clouds has been measured experimentally and it appears that its value is close to 0.75 in the visible. This means that the average geometro-optical cosine of scattering angle is close to 0.5 for nonabsorbing snow. The value of g increases for absorbing grains. It is close to 0.97 for strongly absorbing convex ice particles in random orientation.

The extinction coefficient of snow is difficult to measure directly. However, again one may expect that extinction coefficient of light in snow can be modelled in the same way as for ice clouds. Then it follows as discussed above:

$$k_{ext} = \frac{3c_{ice}}{d_{ef}},\qquad(4.1)$$

where d_{ef} is the effective diameter of grains and c_{ice} is the volumetric concentration of snow crystals. Taking into account that often $c_{ice} \approx 1/3$, we find that and snow optical thickness (SOT) $\tau \approx L/d_{ef}$, where L is the geometrical thickness of the snow layer. Therefore, one may expect that just 5–10 layers of snow particles constitute optically thick medium with the optical thickness above 5–10. This is confirmed also by the naked eye observations of thin snow layers on, say, black underlying surfaces. Usually just several cm of snow is enough to make underlying surface invisible.

It follows from the discussion presented in the previous Chapter that radiative transfer characteristics of weakly absorbing snow layers are determined by mainly two parameters, namely: similarity parameter s and the asymptotic flux attenuation coefficient γ. The similarity parameter s can be derived from the measurements of the diffuse snow reflectance r under diffuse illumination conditions in the visible and near–infrared as discussed above:

$$s = -\frac{1}{4}lnr.\qquad(4.2)$$

A similar equation can be derived, if the plane albedo is measured. The value of γ is obtained from the measurements of the light flux in snow (say, at two levels in snow, see Eq. 3.74). The spectra $s(\lambda)$, $\gamma(\lambda)$ determine the spectral behaviour of snow spectral reflectance and transmittance in the visible and near–infrared.

It should be pointed out that several local optical characteristics of snow are related at small values of probability of photon absorption. For instance, it follows:

$$s = \sqrt{\frac{\beta}{3(1-g)}} \equiv \frac{k}{3(1-g)} = \frac{\beta}{k} = \frac{k_{abs}}{kk_{ext}} = \frac{k_{abs}}{\gamma} = \frac{\gamma}{3k_{tr}},\qquad(4.3)$$

$$\gamma = kk_{ext}, k = \sqrt{3\beta(1-g)},\qquad(4.4)$$

$$k_{tr} = (1 - g)k_{ext}.$$ (4.5)

This means that the snow transport extinction coefficient can be derived from the pair (s, γ):

$$k_{tr}(\lambda) = \frac{\gamma(\lambda)}{3s(\lambda)}.$$ (4.6)

Also it follows for the snow absorption coefficient:

$$k_{abs}(\lambda) = \gamma(\lambda)s(\lambda).$$ (4.7)

4.1.2 Finite Snow Layers

4.1.2.1 Reflectance and Transmittance Measurements

It has been demonstrated above how the snow optical parameters can be derived from light reflection measurements (for a semi-infinite snow layer) in combination with studies of light fluxes in deep snow layers. A similar information can be derived from the measurements of simultaneous spectral transmittance and reflection of optically thick finite snow layers of a given snow depth. Let us assume that the diffuse reflection and transmission coefficients of snow are measured. They are defined by the following equations for the snow layer with the geometrical thickness L as discussed in the previous Chapter:

$$r_s = \frac{shx}{sh(x + by)}, \quad t_s = \frac{shy}{sh(x + by)}.$$ (4.8)

where $x = \gamma L$, $y = 4s$. To simplify the analytical solution of the inverse problem we ignore the difference of the parameter b from unity. The pair (x, y) can be derived from (r, t) given by Eq. (4.8) at $b = 1$ analytically. Let us show it. First of all, we note that

$$\frac{t}{r} = \frac{shy}{shx},$$ (4.9)

where we removed the subscripts to simplify the equations. Also it follows that

$$sh(x + y) = shxchy + chxshy$$ (4.10)

and, therefore,

$$(shx\,chy + chx\,shy)t = shy,$$ (4.11)

$$\left(\frac{r}{t}chy\,shy + \sqrt{1 + sh^2x}\,shy\right)t = shy,\tag{4.12}$$

$$rchy + t\sqrt{1 + sh^2x} = 1,\tag{4.13}$$

$$rchy + \sqrt{t^2 - r^2 + r^2ch^2y} = 1,\tag{4.14}$$

$$chy = \frac{1 - t^2 + r^2}{2r}.\tag{4.15}$$

Taking into account that

$$shx = \frac{r}{t}shy,\tag{4.16}$$

we derive:

$$shx = \frac{r}{t}\sqrt{\frac{(1 + r^2 - t^2)^2}{4r^2} - 1}.\tag{4.17}$$

Therefore, it follows finally:

$$x = arsh\left[\frac{r}{t}\sqrt{\frac{(1 + r^2 - t^2)^2}{4r^2} - 1}\right], y = arcch\left[\frac{1 - t^2 + r^2}{2r}\right].\tag{4.18}$$

Alternatively, we can write (Zege et al. 1980):

$$x = \ln\left[\frac{b}{2t} + \sqrt{1 + \frac{b^2}{4t^2}}\right], y = \ln\left[\frac{b}{2r} + \sqrt{1 + \frac{b^2}{4r^2}}\right],\tag{4.19}$$

where

$$b = \sqrt{(1 + r^2 - t^2)^2 - 4r^2}\tag{4.20}$$

and we used the equality:

$$arshA = \ln(A + \sqrt{1 + A^2}).\tag{4.21}$$

We can derive the pair (γ, s):

$$\gamma = \frac{x}{l}, s = \frac{y}{4}\tag{4.22}$$

and

$$k_{abs} = \gamma y/4, \quad k_{tr} = \frac{4\gamma}{3y}. \tag{4.23}$$

4.1.2.2 Reflectance Measurements of Snow At Two Values of Geometrical Thickness

The pair (γ, y) can be also derived from just reflectance or transmittance measurements at two (or more) values of the geometrical thickness of a snow layer. Let us show it.

The snow spherical albedo can be presented in the following form:

$$r_s = r_{s\infty} - t \exp(-x - y) \tag{4.24}$$

where

$$t = \frac{(1 - \exp(-2y)) \exp(-x)}{1 - \exp(-2y - 2x)}. \tag{4.25}$$

Then measuring the snow reflectance for a semi-infinite snow layer (defined as a layer which does not change with the further increase of snow thickness), we derive

$$y = \ln\left(r_s^{-1}\right). \tag{4.26}$$

The dependence of the parameter $y = 4\,s$ on r_s is given in Fig. 4.1.

Performing measurements of the snow reflectance (with black underlying surface at a fixed snow geometrical thickness h), we derive:

$$\exp(-2\gamma L) \equiv Z = \frac{1 - r_s(h)r_{s\infty}^{-1}}{1 - r_s(h)r_{s\infty}} \tag{4.27}$$

and, therefore,

$$\gamma = \frac{\ln Z^{-1}}{2L}. \tag{4.28}$$

If it is needed, the measurements can be performed at two arbitrary geometrical thicknesses (e.g., not for a semi-infinite snow). Then the pair (γ, y) can be also derived.

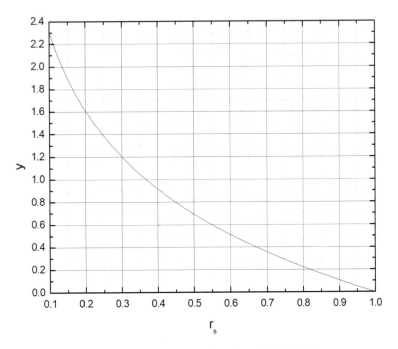

Fig. 4.1 The dependence of the retrieved value of y on the spherical albedo r_s

4.1.2.3 Transmittance Measurements of Snow At Two Values of Geometrical Thickness

The parameters (γ, y) can be also derived from snow transmittance measurements at two snow thicknesses h_1, h_2:

$$t_1 = \frac{(1 - \exp(-2y))\exp(-x_1)}{1 - \exp(-2y - 2x_1)}, t_2 = \frac{(1 - \exp(-2y))\exp(-x_2)}{1 - \exp(-2y - 2x_2)}. \qquad (4.29)$$

Namely, it follows:

$$e^{-2y} = \frac{1 - t_1 e^{x_1}}{1 - t_1 e^{-x_1}} = \frac{1 - t_2 e^{x_2}}{1 - t_2 e^{-x_2}}. \qquad (4.30)$$

It means that

$$(1 - t_1 e^{x_1})(1 - t_2 e^{-x_2}) - (1 - t_2 e^{x_2})(1 - t_1 e^{-x_1}) = 0 \qquad (4.31)$$

or, therefore,

$$t_1 t_2 sh(\gamma(\xi - 1)z) - t_1 sh(\gamma z) + t_2 sh(\xi \gamma z) = 0, \qquad (4.32)$$

where $z = h_1, \xi = h_2/h_1$. This transcendent equation can be used to derive γ. Then it follows:

$$y = \frac{1}{2} \ln \left[\frac{1 - t_1 e^{-yz}}{1 - t_1 e^{yz}} \right]. \tag{4.33}$$

Assuming that $\xi = 2$, one derives:

$$t_1 t_2 - t_1 + 2t_2 ch(\gamma z) = 0 \tag{4.34}$$

or

$$\gamma = \frac{arch(\Upsilon)}{z}, \tag{4.35}$$

where

$$\Upsilon = \frac{t_1 - t_1 t_2}{2t_2}. \tag{4.36}$$

4.1.2.4 Nonabsorbing Snow Layers

In the visible, clean snow is nonabsorbing and, therefore, $k_{abs} = 0$ and k_{tr} can be derived from the transmittance measurements at the snow sample of thickness h. Namely, it follows:

$$k_{tr} = \frac{4}{3L} (T_0^{-1}(\xi, \eta, h) u_0(\xi) u_0(\eta) - \alpha), \tag{4.37}$$

where $T_0^{-1}(\xi, \eta, h)$ is the inverse snow transmission function.

Let us assume that the snow diffuse transmission coefficient $t(\xi, h)$ is measured. Then it follows:

$$k_{tr} = \frac{4}{3L} (t^{-1}(\xi, h) u_0(\xi) - \alpha). \tag{4.38}$$

The equation is even simpler if the diffuse transmission under diffuse illumination (t) is measured:

$$k_{tr} = \frac{4}{3L} (t^{-1}(h) - \alpha). \tag{4.39}$$

4.2 Snow Grain Size Retrieval

Snow is composed of irregularly shaped ice crystals of various forms. As a matter of fact, each crystal is unique and has its own specific shape and geometrical dimensions. Therefore, the notion of snow grain size is not well defined for snow. For the size of particles, one may choose, e.g., the average maximal/minimal dimension of crystals, which can be measured using a lens. Also more advanced approaches to characterize crystals are available. Reflectance spectroscopy makes it possible to determine the effective absorption length (EAL) in snow. This parameter can be used to characterize the snow by optical methods. In particular, it follows for the spherical albedo of a semi-infinite snow layers as $\omega_0 \to 1$:

$$r_s(\lambda) = \exp\left(-\sqrt{\alpha(\lambda)\ell}\right), \tag{4.40}$$

where $\alpha(\lambda)$ is the bulk ice absorption coefficient and ℓ is the effective absorption length. One can see that ℓ can be easily derived from the measurements of $r_s(\lambda)$ at a particular wavelength λ:

$$\ell = \frac{ln^2 r_s}{\alpha(\lambda)}. \tag{4.41}$$

This equation is valid only for clean snow and must be used in the near–infrared, where r_s deviates from 1 (say, at 1020 nm) and the theory used to derive this equation remains valid. The value of ℓ is an important parameter characterizing any snowpack. It can be related to the size of grains using the following equation:

$$d_{ef} = \varsigma L \tag{4.42}$$

where (see Eqs. (3.119), (3.120))

$$\varsigma = \frac{9(1-g)}{16B} \tag{4.43}$$

The value of ℓ can be also found from plane albedo measurements:

$$\ell = \frac{ln^2 r_p}{\alpha(\lambda)u_0^2(\mu_0)} \tag{4.44}$$

and also from the reflectance measurements:

$$\ell = \frac{ln^2(R/R_0)}{\alpha(\lambda)f^2}, \tag{4.45}$$

where R_0 is the nonabsorbing snow reflectance (in the visible). For a vertically homogeneous snow, the derived parameters ℓ and d_{ef} shall not depend on the wavelength. However, due to different penetration of radiation with different wavelength to snowpack, both d_{ef} and ℓ depend on the wavelength in case of vertically inhomogeneous layers. This feature can be used to study the vertical profiles of correspondent parameters. For better understanding of snow vertical inhomogeneity and buried polluted snow layers, one needs to study the snow profiles along the vertical snow walls.

The snow grain diameter can be also retrieved using measurements not at a single wavelength but also in the ice absorption bands centered at specific wavelengths. The spectral observations of the snow surface can be used to assess the vertical distribution of snow grains in the snow top layer owing to the different penetration length of radiation depending on the wavelength.

4.3 Determination of Snow Specific Surface Area

Specific surface area is the property of solids defined as the total surface area of material per unit of mass. Therefore, it follows in the case of snowpack:

$$\Sigma = \frac{N\overline{S}}{N\overline{M}}, \tag{4.46}$$

where N is the number of ice grains in the unit volume, $\overline{S}$ is their average surface area and

$$\overline{M} = \rho_i \overline{V} \tag{4.47}$$

is their average mass, $\rho_i = 0.917 \, g/cm^3$ is the ice density and $\overline{V}$ is the average volume of grains. Therefore, it follows for the snow specific surface area:

$$\sigma = \frac{\overline{S}}{\rho_i \overline{V}}. \tag{4.48}$$

Let us introduce the effective grain diameter using the following equation:

$$d_{ef} = 6\frac{\overline{V}}{\overline{S}}. \tag{4.49}$$

In the case of monodispersed spheres this diameter coincides with the actual diameter of spherical particles. Then it follows:

$$\sigma = \frac{6}{\rho_i d_{ef}}. \tag{4.50}$$

Therefore, one can see that the snow specific surface area can be derived if the average grain diameter is known and vice versa. As it has been discussed above, the measurement of the effective grain size for snow is a difficult matter. Therefore, usually direct measurement techniques are used to estimate SSA. In particular, the methane adsorption (proportional to $\overline{S}$) technique is used to estimate $\overline{S}$. The mass of a snow sample can be found by weighting. Then one derives SSA as the ratio shown in Eq. (4.46).

This is quite an involved procedure, therefore, it looks as an attractive solution of the problem to develop the technique for the determination of SSA using the snow effective absorption length derived from snow reflectance measurements. It has been demonstrated that ℓ can be presented in the following form:

$$\ell = \frac{32B\overline{V}}{3(1-g)\overline{\Phi}}, \tag{4.51}$$

where $\overline{\Phi}$ is the average surface area of grains, g is the asymmetry parameter, and B is the absorption enhancement parameter. The accuracy of this expression is high for weakly absorbing snow (for the wavelengths below 1000 nm or so). It also follows:

$$\ell = \frac{32B}{3(1-g)\rho_i\sigma} \tag{4.52}$$

or

$$\sigma = \frac{32B}{3(1-g)\rho_i\ell}. \tag{4.53}$$

Equation (4.53) can be presented in the following general form:

$$\sigma = \frac{A}{\rho_i\ell}, \tag{4.54}$$

where

$$A = \frac{32B}{3(1-g)} \tag{4.55}$$

depends on the snow type. It follows from Eq. (4.54):

$$\ln \sigma = \varepsilon - \ln h, \tag{4.56}$$

where

$$h = \rho_i \ell \qquad (4.57)$$

and

$$\varepsilon = \ln A. \qquad (4.58)$$

It is essential to study the value of A for different types of snow experimentally.

4.4 Determination of Snow Impurity Content

4.4.1 General Equations

Snow originates from the atmosphere, where various aerosol (solid and water soluble) particles are present. Therefore, it contains not only frozen water but also other chemical substances such as brown/black carbon, dust, litter from the trees, etc. Also various types of algae in various quantities depending on location can be present. Reflectance spectroscopy is traditionally used to understand the chemical composition of various substances. In this section we consider the application of reflectance spectroscopy to derive the chemical composition and concentration of various impurities in snow. The variation of pollution load in various snow samples can be assessed from Table 4.1.

Let us assume that the diffuse reflectance of snow under diffuse illumination conditions for semi-infinite is studied. Such a scheme makes it possible to avoid the use of auxiliary functions, which are potential sources of an error. The measured spherical albedo can be presented as

$$r_s = \exp(-y). \qquad (4.59)$$

Therefore, the spectral values of the similarity parameter y can be directly retrieved from the measured reflectance spectrum. In particular, it follows:

$$y(\lambda) = \ln\left(r_s^{-1}(\lambda)\right). \qquad (4.60)$$

Therefore, the inverse problem involving multiple light scattering in snow is reduced to the determination of the type of pollutants and their load from the similarity parameter

$$y(\lambda) = 4\sqrt{\frac{\beta}{3(1-g)}}. \qquad (4.61)$$

Table 4.1 The measured concentration of impurities at several locations

Impurity	Concentration	Location	Source
Soot	0.1–0.3 ng/g	South Pole	Flanner et al. (2007)
Soot	1–30 ng/g	Summit, Greenland	Flanner et al. (2007)
Soot	28–210 ng/g	Rural Michigan	Flanner et al. (2007)
Soot	80–826 ng/g	French Alps	Flanner et al. (2007)
Soot	17–5700 ng/g	Urban Michigan	Flanner et al. (2007)
Dust	0.04–0.1 mg/g	European Alps	Biagio et al. (2015)
Dust	20–50 mg/g	Alpine and subalpine	Skiles et al. (2012)
Algae	0.1–1.5 cells/mL	Antarctic penisula	Gray et al. (2020)
Algae	1–25cells/mL	Greenland ice sheet	Stibal et al. (2015)

More precisely, the spectral probability of photon absorption β must be studied because for most of cases (but not always!) light scattering in snow is governed mostly by ice grains and not by impurities. Therefore, we can derive from Eq. (4.61)

$$\beta(\lambda) = \Theta \ln^2(r_s(\lambda)), \qquad (4.62)$$

where the parameter $\Theta = 3(1 - g)/4$ can be derived assuming the asymmetry parameter for ice grains. The error in the estimated Θ leads to the uncertainties in the derived PPA and also concentration of impurities. In particular, we may assume that $g = 3/4$ in the visible and near–infrared and, therefore, $\Theta = 3/16$.

Now we need to relate the PPA $\beta(\lambda)$ to the concentration of pollutants. This can be done in the following way. Let us assume that pollutants are externally mixed with snow grains and their influences on scattering (and light extinction) in snow is negligible as compared to the contribution of snow grains. Then it follows:

$$\beta(\lambda) = \frac{k_{abs}^{ice}(\lambda) + k_{abs}^{imp}(\lambda)}{k_{ext}^{ice}}, \qquad (4.63)$$

where

$$k_{ext}^{ice} = \frac{3c_{ice}}{d_{ef}}, \sigma_{abs}^{ice}(\lambda) = Bc_{ice}\alpha_{ice}(\lambda), k_{abs}^{imp}(\lambda) = c_{pol}K_{pol}(\lambda). \qquad (4.64)$$

Here, c_{ice} is the volumetric concentration of ice in snow, c_{pol} is the volumetric concentration of pollutants in snow, d_{ef} is the effective grain diameter, $\alpha_{ice}(\lambda)$ is the bulk ice absorption coefficient, B is the absorption enhancement factor, and $K_{pol}(\lambda)$ is the volumetric absorption coefficient of pollutants:

$$K_{pol}(\lambda) = \frac{\overline{C}_{abs}^{pol}}{\overline{V}_{pol}}, \tag{4.65}$$

where $\overline{C}_{abs}^{pol}$ is the average absorption cross section of pollutants and $\overline{V}_{pol}$ is the average volume of impurity particles. One may also introduce mass absorption coefficient of pollutants:

$$K_{pol,mass}(\lambda) = \frac{K_{pol}}{\rho_{pol}}, \tag{4.66}$$

where ρ_{pol} is the density of a pollutant. Therefore, it follows for the probability of photon absorption:

$$\beta(\lambda) = \left[B\alpha_{ice}(\lambda) + cK_{pol}(\lambda) \right] \frac{d_{ef}}{3}, \tag{4.67}$$

where $c = c_{pol}/c_{ice}$ is the relative concentration of pollutants. One can see that the value of c can be derived from PPA:

$$c = \left[\frac{3\beta}{d_{ef}} - B\alpha_{ice} \right] \frac{1}{K_{pol}} \tag{4.68}$$

or

$$c = \frac{3\beta}{K_{pol}d_{ef}}, \tag{4.69}$$

if the wavelength selected for the inversion procedure is in the visible, where one can neglect the second term in Eq. (4.68) due to small light absorption by ice in the visible. One can see that the relative concentration of pollutants can be derived if the parameters β, d_{ef}, and K_{pol} (or alternatively, $K_{pol,mass}$) are known. The pair (β, d_{ef}) in Eq. (4.69) can be derived as discussed above. The main problem in the precise determination of c is the determination of the volumetric absorption coefficient of pollutants (see Eq. 4.69). The value of K_{pol} depends on the size distribution $f(a)$ of pollutant particles and also on their complex refractive index $m_{pol} = n_{pol} - i\chi_{pol}$. This makes retrievals of c problematic because both $f(a)$ and m_{pol} (and, therefore, K_{pol}) are not known in advance. Let us consider this problem for main snow pollutants such as soot and dust.

4.4.2   Soot

In the case of soot, the scatterers are much smaller than the wavelength of incident light and, therefore,

$$\overline{C}_{abs,pol}(\lambda) = \Psi \alpha_{pol}(\lambda)\overline{V}_{pol}, \tag{4.70}$$

$$K_{pol}(\lambda) = \Psi \alpha_{pol}(\lambda), \quad \alpha_{pol}(\lambda) = \frac{4\pi \chi_{pol}}{\lambda}, \tag{4.71}$$

where

$$\Psi = \frac{9n_{pol}}{\left(n_{pol}^2 + 1 - \chi_{pol}^2\right)^2 + 4n_{pol}^2 \chi_{pol}^2}. \tag{4.72}$$

There is some uncertainty with respect to the spectral refractive index of soot. Assuming that $n_{pol} = 1.75$, $\chi_{pol} = 0.45$, we derive that $\Psi = 0.9$. Finally, it follows from Eq. (4.69) for the relative concentration of soot in snow:

$$c = \Pi \frac{\beta}{\alpha_{pol}d_{ef}}, \tag{4.73}$$

where $\Pi = 3/\Psi$. As a matter of fact, some of soot can be internally mixed. Therefore, the concentration as derived from this equation can be biased.

4.4.3   Dust

The derivation of the pollution load for the case of dust is even more complicated because: (1) dust can influence both light scattering and absorption processes in snow; (2) the spectral refractive index of dust differs depending on the origin of dust.
   The spherical albedo of dust–loaded snow can be presented as

$$r(\lambda) = \exp\left(-\varepsilon \sqrt{\left(c_{ice} B \alpha_{ice}(\lambda) + c_{pol} k_0 \left(\frac{\lambda}{\lambda_0}\right)^{-\nu}\right) l_{tr}}\right), \tag{4.74}$$

where $l_{tr} = 1/(1-g)k_{ext}$ is the transport extinction length, $\varepsilon = 4/\sqrt{3}$, k_0 is the volumetric absorption coefficient of impurities at the wavelength λ_0.
   It follows from Eq. (4.74):

$$r(\lambda) = \exp\left(-\sqrt{\left(\alpha_{ice}(\lambda) + q\left(\frac{\lambda}{\lambda_0}\right)^{-\nu}\right)}\ell\right),$$ (4.75)

where

$$\ell = c_{ice} B \varepsilon^2 l_{tr}, q = \frac{k_0}{B} c,$$ (4.76)

$c = c_{pol}/c_{ice}$ is the relative concentration of pollutants. It follows that the spectrum of the dust loaded snow depends on three parameters: ν, q, ℓ. They can be derived from the measurements of spherical albedo at three wavelengths. Namely it follows from Eq. (4.75):

$$\nu = \frac{2lnz}{ln\left(\frac{\lambda_1}{\lambda_2}\right)}, q = \frac{b}{l}, \ell = \frac{\{ln^2 r(\lambda_3) - b(\lambda_3/\lambda_0)^{-\nu}\}}{\alpha_{ice}(\lambda_3)},$$ (4.77)

where

$$b = ln^2(r(\lambda_1))\left(\frac{\lambda_1}{\lambda_0}\right)^{\nu}, z = \frac{lnr(\lambda_2)}{lnr(\lambda_1)}.$$ (4.78)

Equation (4.77) can be used to derive the snow grain size of polluted snow and also the concentration of pollutants c. Namely, it follows from Eq. (4.76):

$$d_{ef} = \frac{9(1-g)\ell}{16B}, c = \frac{B}{k_0}q.$$ (4.79)

One can assume that $\frac{B}{1-g} = 9$ and $B = 1.6$ (Kokhanovsky et al. 2021a). This makes it possible to estimate the snow grain size from the measurements of *EAL*. For the determination of the concentration of pollutants we need to know the value of k_0. Clearly, the value of k_0 is correlated with the value of the absorption Angstrom parameter ν. The following correlation equation derived from the Mie theory under certain assumptions with respect to the size distribution of dust grains and also their complex refractive index can be used (Kokhanovsky et al. 2021a):

$$k_0 = \sum_{i=0}^{2} b_i \nu^i,$$ (4.80)

where $b_0 = 10.916, b_1 = -2.0831, b_2 = 0.5441$ and the volumetric absorption coefficient is expressed in 1/mm.

The accuracy of the theory developed above is demonstrated in Fig. 4.1, where the following triplet of the wavelength has been used to retrieve ν, q, ℓ from the measured plane albedo spectra: 410, 500, and 825 nm. One can see that three-parameter formula

(4.75) can be used to determine the snow plane albedo

$$r_p = r^{u_0(\mu_0)}. \qquad (4.81)$$

The various parameters of snow derived from the spectra shown in Fig. 4.2 are presented in Table 4.2. The relative mass concentration is found as follows:

$$c_m = \frac{\rho_d}{\rho_i} c, \qquad (4.82)$$

where $\rho_d = 2.65 g/cm^3$ is the density of dust and $\rho_i = 0.917 g/cm^3$ is the density of ice.

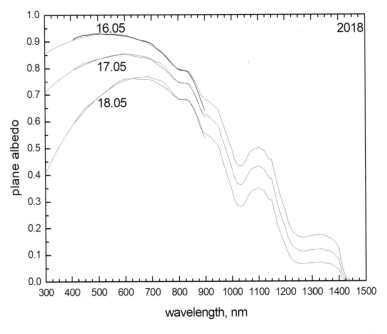

Fig. 4.2 The spectral snow albedo: measurements–red lines, calculations using Eq. (24)—blue lines. Dates refer to the acquisitions of the albedo spectra (Kokhanovsky et al. 2021a)

Table 4.2 The parameters of dust–loaded snow derived at three dates at the Torgnon site (Kokhanovsky et al. 2021a)

Day	SZA, degrees	α	q, 1/mm	ℓ, mm	k_0, 1/mm	d, mm	c_m, ppm
16.05.2018	24.44	3.00	2.391e-5	18.40	9.63	1.15	11.7
17.05.2018	27.21	2.51	1.517e-4	25.60	9.11	1.60	77.4
18.05.2018	26.98	3.36	2.304e-4	37.28	10.11	2.33	106.9

The formulae given above are valid only for external mixture of pollutants and ice grains. In practice, part of particles can be externally mixed and another part of impurity particles can be contained inside (internal mixture) ice grains (increasing absorption), which complicates the retrieval procedure. In addition, several and not just one pollutant (including algae) may present in snow making retrievals even more involved.

4.5 Spaceborne Remote Sensing of Snow

4.5.1 Spaceborne Instrumentation

Snow covers huge polar regions and also it has considerable extent in the Northern hemisphere during winter. Snow permanently presents almost over whole Antarctica, which is the fifth-largest continent almost twice of Australia ($14,200,000\ km^2$). The same is true for Greenland ($2,166,086\ km^2$). Therefore, it is of importance to develop techniques for snow monitoring using satellite instrumentation. Usually passive optical (visible and near – infrared) and microwave sensors are used for snow monitoring from space. In particular, snow water equivalent can be derived from passive microwave observations. In this chapter we consider only the optical snow remote sensing. The list of optical instruments most often used for snow monitoring is given in Table 4.3. They are also briefly discussed below.

Multiple Spectral Imager (MSI) on board Sentinel-2 operates in a broad spectral range 0.443–2.2 μm providing measurements at a high spatial resolution (10–60 m, depending on the channel). The 12 detectors on each focal plane are mounted in a staggered formation to cover the whole 20.6° instrument field of view, resulting in a compound swath width of 290 km on the ground track.

Ocean and Land Color Instrument (OLCI) is a single-view push-broom imaging spectrometer, which measures the top-of-atmosphere reflected light in the spectral range 400-1020 nm with a spatial resolution of 300 m. It operates on board Sentinel-3. The swath is 1270 km.

Sea and Land Surface Temperature Radiometer (SLSTR) on board Sentinel-3 uses two independent scan mirrors rotating in opposite directions each scanning at a rate of 200 scans per minute providing dual-view capability. The swath width is 1400 km for the nadir observation and 700 km for the along track view. The spatial resolution is 500 m in solar channels. It performs measurements in visible, short/mid wave infrared. It also has thermal channels located at 10.85 and 12.02 μm.

Second Generation Global Imager (S-GLI) is a push-broom instrument providing 11 channels in the visible and near infrared spectral region and 6 channels in the shortwave infrared and thermal infrared regions of electromagnetic spectrum. Two channels have a 3-view capability and perform measurements of the polarization state of reflected light, which is crucial for atmospheric correction procedures. The spatial resolution is 0.25–1 km depending on the channel.

Table 4.3 Satellite optical instrumentation (the most important channels used for snow monitoring are underlined)

Device	Channels (nm)	Spatial sampling, m	Swath/global coverage/revisit time	Spacecraft/launch date	Observation mode
MSI	442.7, 492.4, 559.8, 664.6, 704.1, 740.5, 782.8, 832.8, 864.7, 945.1, 1373.5, 1613.7, 2202.4 nm	10/20/60 (depending on the channel)	290 km/10 days at the equator revisit time	S-2A/23.06.2015 S-2B/07.03.2017	Single view multispectral instrument The specification of channels is given for S-2A The channels for S-2B are slightly shifted
OLCI	400, 412.5, 442.5, 490, 510, 560, 620, 665, 673.75, 681.25, 708.75, 753.75, 761.25, 764.375, 767.5, 773.75, 778.75, 865, 885, 900, 940, 1020 nm	300	1270 km/ 4 days	S-3A/16.02.2016 S-3B/25.04.2018 S-3C/2021 (planned) S3D/202x (planned)	Single-view push-broom imaging spectrometer
SLSTR	VIS: 554, 659, 868 nm SWIR: 1.374, 1.613, 2.25 μm, MWIR-TIR: 3.742, 10.85, 12.02 μm	500–1000 (depending on the channel)	1400 km /nadir view/1 day 740 km/ oblique view/2 days	The same as for OLCI	Dual-view radiometer The MWIR-TIR channels have spatial resolution of 1 km. Otherwise, it is 0.5 km

(continued)

Table 4.3 (continued)

Device	Channels (nm)	Spatial sampling, m	Swath/global coverage/revisit time	Spacecraft/launch date	Observation mode
S-GLI	19 channels: 0.38–12 µm (0.38,0.412,0.443,0.49,0.53,0.565,0.6735,0.763,0.8685,1.05,1.38,1.63,2.21,10.8,12 µm)	250/1000 (depending on the channel)	1150 km (VNR) 1400 km (IRS) Global observation: every 2 or 3 days	23.12.2017	A multi-band optical radiometer. It contains of two sensors: SGLI-VNR (an electronic scan) and SGLI-IRS (a mechanical scan). The polarization of reflected light is measured at 673.5 and 868.5 nm (multi-angular observations)
VIIRS	22 spectral bands 0.4–12.01 µm (0.412, 0.445,0.488,0.555,0.64,0.672,0.746,0.865,0.7,1.24,1.38,1.61,2.25,3.7,3.74,4.05,8.55,10.76,11.45, 12.01 µm channels)	375/750 m	3040 km (cross track), each granule covers the area 3040 km × 570 km (48 scans/85 s)	28.10.2011	Single-view scanning radiometer (a rotating mirror reflects radiation onto a set of CCD detectors)
MODIS	36 spectral bands 0.41–14.3 µm (250 m: 0.659, 0.865 µm, 500 m: 0.47,0.555,1.24,1.64, 2.13 µm, 1 km: 0.412,0.443,0.488,0.531,0.551,0.667, 0.678, 0.748,0.869,0.905,0.936,0.94,1.375 µm 1 km spatial resolution thermal emissive bands: 3.75,3.96,4.05,4.47,4.52,6.72,7.33,8.55,9.73,11.03,12.02,13.34,13.64,13.94, 14.24 µm	0.25-1 km, depending on the channel	2330 km (cross track) by 10 km (along track), 10:30a.m (Terra) 13:30p.m (Aqua)	18.12.1999 (Terra) 04.05.2002 (Aqua)	Single-view imaging spectro-radiometer

Visible Infrared Imaging Radiometer Suite (VIIRS) observes and collects global satellite observations that span the visible and infrared wavelengths across land, ocean, and atmosphere. A whiskbroom radiometer by design, it has 22 channels ranging from 0.41 μm to 12.01 μm. Five of these channels are high-resolution image bands or L-bands, and sixteen serve as moderate-resolution bands or M-bands. VIIRS is one of five instruments onboard the Suomi National Polar-orbiting Partnership (NPP) satellite platform that was launched on October 28, 2011. SNPP (formerly called the National Polar-orbiting Operational Environmental Satellite System (NPOESS) Preparatory Project) serves as a bridge between the Earth Observing System (EOS) satellites and the next-generation NASA-NOAA Joint Polar Satellite System (JPSS). Four JPSS missions have been planned to last through 2031, and each of them will host a VIIRS instrument as part of their payload. JPSS-1 (NOAA-20) was launched on November 18, 2017. The swath is 3000 km with nearly global coverage every day (1:30p.m. local solar time).

The Moderate Resolution Imaging Spectroradiometer (MODIS) was launched into Earth orbit by NASA in 1999 on board of Terra (EOS AM) and in 2002 on board the Aqua (EOS PM) satellites. The instruments capture data in 36 spectral bands from 0.4 to 14.4 microns at varying spatial resolutions (2 bands at 250 m, 5 bands at 500 m, and 29 bands at 1 km). Together these instruments image the entire Earth every 1–2 days.

4.5.2 Cloud Screening

The retrieval of snow properties using spaceborne observations requires the application of cloud screening and atmospheric correction algorithms. The cloud screening over dark ocean surfaces could be easily performed because clouds are much brighter as compared to underlying ocean surface. The snow and cloud surfaces are both bright. Therefore, the cloud screening over snow is a complex matter. Several tests are used to select the cloudless satellite pixels. They are summarized in Table 4.4. The tests are based on the fact that the reflected solar radiation at the top-of-atmosphere for scenes with snow and clouds have different spectral, directional, and polarization characteristics. These differences are explored to distinguish underlying snow and cloud surfaces. The most important feature used for cloud identification over bright surfaces is screening of gaseous atmosphere underneath the clouds for cloudy scenes. This leads to the increase of top-of-atmosphere reflectance in the absorption bands of gases with maxima of concentrations close to the terrestrial surface such as oxygen and water vapor. The threshold values (THV) used to distinguish the cloud scenes depend on the underlying surface height, geometry of observation, and channel. They can be found preparing the histogram of TOA reflectances at specific gaseous absorption wavelengths for clear and cloudy scenes. Such histograms show clear separation of two regimes (cloudy, clean), which can be used to the determination of

Table 4.4 Cloud tests based on the value of reflectances R at several spectral regions and observation geometries. The value of Δ is the standard deviation of reflectances measured at a given point at different times or at a given time for an area surrounding the location of interest. The tests 6 and 7 refer to the reflectances at the scattering angle θ defined after Eq.(3.32)

N	Name	Test	Physical property
1	Oxygen A-band test	R(761 nm) > THV1	Larger reflectance for cloudy scenes in the gaseous absorption bands due to screening of gases under the clouds
2	Water vapor absorption test	R(940 nm) > THV2 R(1340 nm) > THV3	See above
3	Spectral reflectance check	R(1020 nm) < THV4 R(1640 nm) < THV5	Smaller absorption by particles in clouds as compared to crystals in snow due to their smaller sizes
4	Spatial inhomogeneity check	Δ > THV6	High spatial variability of cloudy scenes
5	Temporal inhomogeneity check	Δ > THV7	High temporal variability of ground scenes covered by clouds
6	Angular reflectance check	R(148°)/R(120°) > THV8	Higher variability of angular distribution of reflected radiation for clouds (e.g., rainbow, glories)
7	Polarization check	R(148°)/R(120°) > THV9	See above except for the degree of polarization P

THVs and the development of corresponding cloud screening procedures. Another possibility is to use the TOA reflectance in the absorption bands of water and ice. Water droplets and ice crystals in clouds are smaller as compared to the size of snow grains, therefore, the respective TOA reflectances are larger for the cloudy scenes. The THVs can be determined in the same way as described above. Also the liquid water clouds can be identified by the shift of the absorption band as compared to the case of ice grains in snow (see Appendix). In addition water clouds have unique directional and polarization characteristics (rainbow, glory) not existing for snow fields. Terrestrial surfaces have unique features for a given location. The existence of these features in the satellite images points to the absence of clouds. In addition, one can check satellite images at a given location with frequent re-visit time. The cloudy pixels are characterized by larger temporal variability as compared to clean scenes. All features underlined above can be used to prepare a cloud mask and select cloud free scenes for the subsequent snow property retrievals.

4.5.3 Atmospheric Correction

The optical signals detected by the satellite instrumentation are influenced not only by the underlying snow surface. The characteristics of the propagation channel (atmosphere) is of importance as well. Therefore, the spectral and angular distribution of Stokes parameters of reflected solar light at the top-of-atmosphere must be measured. This makes it possible to determine both atmosphere and underlying surface properties for a given satellite scene. In most of cases the information content of measurements is low and one should use additional databases or assumptions to make the determination of snow characteristics from spaceborne measurements.

Often the spectral measurements of the TOA reflectance $R(\lambda)$ are performed. Then assuming that the underlying surface is Lambertian, one can use the following relationship between $R(\lambda)$ and albedo of underlying surface $r_s(\lambda)$ (Liou 1992):

$$R = R_a + \frac{r_s(\lambda)T_a(\lambda)}{r_s(\lambda)r_a(\lambda)}, \tag{4.83}$$

where T_a is the atmospheric transmittance on the way from the TOA to the surface and back to the satellite position, R_a is the atmospheric reflectance for the case of black underlying surface and r_a is the spherical albedo of atmosphere. These functions depend on the aerosol/molecular optical thickness, phase function, and atmospheric single scattering albedo. The spectral spherical albedo of underlying surface can be derived from Eq. (4.83) analytically:

$$r_s(\lambda) = \frac{R(\lambda) - R_a(\lambda)}{T_a(\lambda) + (R(\lambda) - R_a(\lambda))r_a(\lambda)}. \tag{4.84}$$

Let us assume that the atmospheric optical thickness is equal to zero. Then it follows: $R_a(\lambda) = r_a(\lambda) = 0, T_a(\lambda) = 1$, and $r_s = R(\lambda)$ as it should be in the assumption of underlying Lambertian surface. If the atmospheric optical thickness is not zero, the functions $R_a(\lambda)$, $T_a(\lambda)$, and $r_a(\lambda)$ can be estimated using the radiative transfer modelling (see Appendix).

4.5.4 Snow Albedo, Snow Grain Size and Snow Specific Area

Yet another approach for atmospheric correction is the use of near-infrared channels. In this case the influence of the propagation channel is often can be neglected and one can assume that the satellite measures directly the snow reflectance function, which can be presented as (see Eqs. 2.101, 3.109, 4.42):

$$R = R_0 \exp(-u(\mu_0)u(\mu)\sqrt{\alpha\ell}). \tag{4.85}$$

In particular, if one uses the channels $\lambda_1 = 865$ and $\lambda_2 = 1020$ nm, one can estimate the parameters R_0 and ℓ in general expression (4.85) from reflectance measurements. Namely, Eq. (4.85) can be used for the value of ℓ (at 1020 nm) and $R_0 = R_1^{\zeta} R_2^{1-\zeta}$, $\zeta = 1/(1-b)$, $b = \sqrt{\alpha_1/\alpha_2}$, where indices signify the wavelengths used. This makes it possible to determine the snow reflection function at all channels including visible and UV wavelengths using Eq. (4.85) and the derived parameters R_0 and ℓ. In this way one can determine not only spectral and broadband snow albedo but also one can derive snow grain size and snow specific surface area using the approach described for the case of ground observations at the beginning of this Chapter. The typical OLCI spectrum over snow is given in Fig. 4.3. The channel located at 1020 nm is sensitive to the size of crystals in snow. The retrieval results based on OLCI measurements are shown in Fig. 4.4.

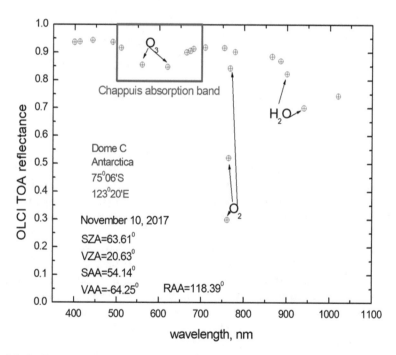

Fig. 4.3 OLCI TOA spectrum over snow at solar zenith angle (SZA) equal to 63.61°, viewing zenith angle (VZA) equal to 20.63°, solar azimuthal angle (SAA) 54.14 degrees, and viewing azimuthal angle (VAA) of −64.25°. The relative azimuthal angle is equal to 118.39 degrees. The OLCI measurements have been performed over Dome C (Antarctica) on November 10, 2017. The regions affected by various atmospheric gases are shown (Kokhanovsky et al. 2021b)

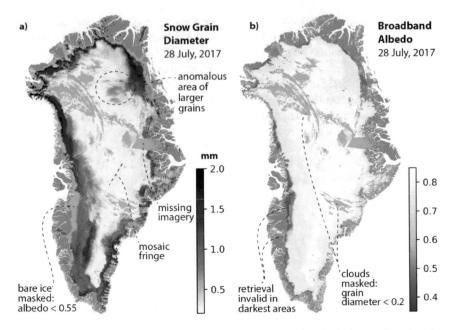

Fig. 4.4 Sentinel-3 daily mosaic of **a** snow grain size and **b** broadband albedo over Greenland for July 28, 2017 (Kokhanovsky et al. 2019)

In addition, the difference between measured TOA reflectance (see Fig. 4.3) and calculated snow reflectances in the visible and UV can be used to determine the aerosol optical thickness and also total ozone. Such an approach is valid only for clean snow surfaces. Otherwise, the spectrum of reflected light governed not only by the atmospheric effects and properties of snow grains. The absorption and scattering of light by impurities in snow must be accounted as well.

The inter-comparison of snow broadband albedo retrieved from space and measured on ground in the framework SICE algorithm (Kokhanovsky et al. 2020) is shown in Fig. 4.5. One can see that the satellite observations provide similar values of BBA as compared to ground observations. The SICE BBA albedo retrieval procedure relies on the measurements of TOA reflectance at 865 and 1020 nm with subsequent determination of spectral and snow albedo as described above. Also the snow grain size and specific surface area can be retrieved from optical spaceborne measurements. The results of satellite retrievals of SSA are compared to those derived from ground observations are shown in Fig. 4.6. One can see close agreement of satellite-derived and ground–measured of SSA. The monthly grain diameter, snow specific surface area, albedo at 1020 nm and broadband albedo for July 2017 over Greenland as derived from OLCI is shown in Fig. 4.7.

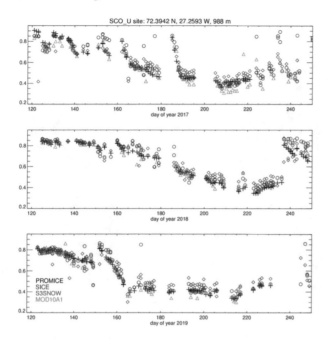

Fig. 4.5 The broadband albedo time series from the SCO_U PROMICE automatic weather station site in comparison to the broad band retrievals from the SICE retrieval, the OLCI retrieval after (Kokhanovsky et al. 2020) (S3SNOW) and for the NASA MODIS MOD10A1 product (Hall and Riggs 2016)

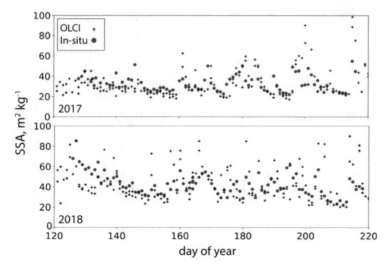

Fig. 4.6 The temporal trends of satellite derived (OLCI) and in situ SSA for 2017 and 2018 years in Greenland (75.625 N, 35.973 W) (Kokhanovsky et al. 2020)

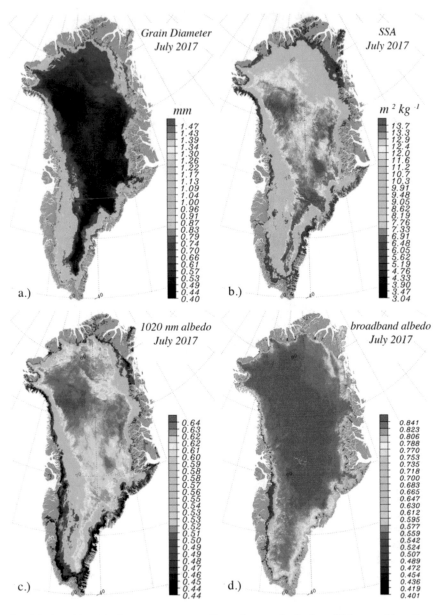

Fig. 4.7 Monthly grain diameter **a** snow specific surface area **b** albedo at 1020 nm and broadband albedo **c** for July 2017 as derived from OLCI (Kokhanovsky et al. 2019)

4.5.5 Snow Fraction and Snow Extent

Snow extent and albedo are main characteristics of snow cover as far as climatic effects of snow are of concern. Snow extent is monitored on a global scale using satellite observations in a broad range of electromagnetic spectrum from UV to microwaves. Microwave instruments can be used for monitoring snow cover under cloud fields. Optical methods can be used for studies of snow cover for clear sky only. The spatial resolution of passive microwave—derived snow cover and snow water equivalent is rather coarse (25 km). This is not the case for optical measurements, where the observations on the scale of several hundred meters are common (depending on the sensor).

Normalized differential snow index is used as a main parameter to distinguish snow free and snow-covered surfaces. It is defined as

$$\text{NDSI} = \frac{R_1 - R_2}{R_1 + R_2},\tag{4.86}$$

where R_1 is the top-of-atmosphere (TOA) reflectance in the visible and R_2 is the TOA reflectance in the near–infrared. Depending on the remote sensing instrument specifications, various bands are selected. In particular, MODIS NDSI relies on the channels located at 0.55 and 1.6 µm. The reflectance R_2 is small over snow. Therefore, pixels with relatively large values of NDSI are assumed to be snow covered (in addition to the large the reflectance values of reflectance at 865 nm). Various threshold values are employed in various algorithms. MODIS snow extent algorithm relies on several spectral tests including NDSI > 0.4, R(650 nm) > 0.1, R(865 nm) > 0.11 to detect snow in a non-densely forested region (Hall et al. 2015).

In some cases, the ground scene is only partially covered by snow. Then one can assume that the reflectance R can be presented as a weighted sum of reflectances of snow (R_s) and underlying free of snow surface (R_c):

$$R(\lambda) = (1 - f)R_c(\lambda) + f R_s(\lambda),\tag{4.87}$$

where f is the snow fraction and it is assumed that the atmospheric correction of the TOA reflectance R has been performed. One can see that the measured and atmospherically corrected reflectance is a linear function of the snow fraction f. It follows from Eq. (4.87):

$$f = \frac{R(\lambda) - R_c(\lambda)}{R_s(\lambda) - R_c(\lambda)}.\tag{4.88}$$

Therefore, the subpixel snow fraction can be estimated, if the snow and underlying free of snow surface reflectances are known. The example of spectral albedos (angular integrals of reflectance) of various snow and underlying free of snow surfaces are shown in Fig. 4.8.

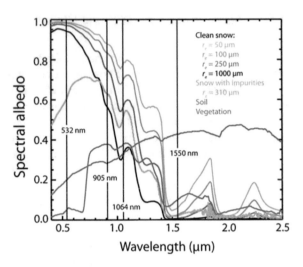

Fig. 4.8 The spectral albedo of soil, vegetation and snow surfaces with various effective grain radii r_e. The case of dust-polluted snow is also shown. The vertical lines correspond to the laser wavelengths used for the snow depth estimation (Deems et al. 2013)

The variability of the clean snow reflectance R_s in the visible is low. Therefore, it can be estimated using radiative transfer calculations. Alternatively, R_s can be taken from the brightest pixels surrounding the pixel under study. The value of R_c can be taken from the same pixel except recorded before snowfall. If the measurements are performed at the near infrared wavelength, where the snow reflectance is low (say, at 2.1 μm), one derives from Eq. (4.87):

$$f = 1 - \frac{R_{nir}}{R_{c,nir}}. \tag{4.89}$$

In this case one needs to know the underlying surface reflectance before the snowfall $R_{c,nir}$. As a matter of fact, the reflectances of many surfaces in the NIR and visible can be often related. For instance, Kaufman and al. (2002) used the linear relationship

$$R_{c,nir} = \gamma R_{c,vis}, \tag{4.90}$$

where $\gamma \approx 2$. Assuming that measurements are performed in the visible (λ_1) and near infrared (λ_2), one derives from Eq. (4.87):

$$R(\lambda_1) = (1 - f)R_c(\lambda_1) + f R_s(\lambda_1), \tag{4.91}$$

$$R(\lambda_2) = (1 - f)R_c(\lambda_2) + f R_s(\lambda_2). \tag{4.92}$$

Assuming that $R_s(\lambda_2) < < 1$ and using Eq. (4.90), one derives from Eqs. (4.91, 4.92):

$$R(\lambda_2) = (1 - f)\gamma R_c(\lambda_1). \tag{4.93}$$

It follows from Eqs. (4.91, 4.93):

$$R(\lambda_1) = R(\lambda_2)/\gamma + f R_s(\lambda_1) \tag{4.94}$$

or

$$f = \frac{R(\lambda_1) - R(\lambda_2)/\gamma}{R_s(\lambda_1)}, \tag{4.95}$$

where $R_s(\lambda_1)$ can be taken from the surrounding brightest pixel, $R(\lambda_1)$ and $R(\lambda_2)$ are measured reflectances at two wavelengths and γ depends on the underlying surface type.

References

Di Mauro, B., B. Fava, L. Ferrero, R. Garzonio, G. Baccolo, B. Delmonte, and R. Colombo. 2015. Mineral dust impact on snow radiative properties in the European Alps combining ground, UAV, and satellite observations. *Journal of Geophysical Research Atmosphere* 120: 6080–6097. https://doi.org/10.1002/2015JD023287.

Deems, J.S., T.H. Painter, and D.C. Finnegan. 2013. Lidar measurement of snow depth: A review. *Journal of Glaciology* 59 (215): 467–479.

Elmes, A., C. Levy, A. Erb, D.K. Hall, T.A. Scambos, N. DiGirolamo, and C. Schaaf. 2021. Consequences of the 2019 Greenland Ice Sheet melt episode on albedo. *Remote Sensing* 13: 227. https://doi.org/10.3390/rs13020227.

Flanner, M.G., C.S. Zender, J.T. Randerson, and P.J. Rasch. 2007. Present - day climate forcing and response from black carbon in snow. *Journal of Geophysical Research* 112: D11202. https://doi.org/10.1029/2006JD008003.

Gray, A., et al. 2020. Remote sensing reveals Antarctic green snow algae as important terrestrial carbon sink. *Nature Communications* 20 (11): 2527. https://doi.org/10.1038/s41467-020-16018-w, www.nature.com/naturecommunication.

Hall, D.K., G.A. Riggs, V.V. Salomonson, N.E. DiGirolamo, and K.J. Bayr. 2012. MODIS snow—cover products. *Remote Sensing of Environment* 83: 181–194.

Hall, D.K., A. Frei, and S.J. Dery. 2015. Remote sensing of snow extent. In *Remote sensing of the cryosphere*. N.Y.: Wiley-Blackwell.

Kaufman, Y.J., R. G. Kleidman, D.K. Hall, J. Vanderlei Martins, and J.S. Barton. 2002. Remote sensing of subpixel snow cover using 0.66 and 2.1 µm channels. *Geophysical Research Letters* 29 (16): 1781. https://doi.org/10.1029/2001GL013580.

Kokhanovsky, A.A. 2004. Scaling constant and its determination from the simultaneous measurements of light reflection and methane adsorption by snow samples. *Optics Letters* 31: 3282–3284.

Kokhanovsky, A.A., M. Lamare, O. Danne, et al. 2019. Retrieval of Snow Properties from the Sentinel-3 Ocean and Land Colour Instrument. *Remote Sensor* 11: 2280. https://doi.org/10.3390/rs11192280.

Kokhanovsky, A.A., J.E. Box, B. Vandecrux, K.D. Mankoff, M. Lamare, A. Smirnov, and M. Kern. 2020. The determination of snow albedo from satellite measurements using fast atmospheric correction technique. *Remote Sensor* 12: 234. https://doi.org/10.3390/rs12020234.

Kokhanovsky, A.A., B. Di Mauro, R. Garzonio, and C. Roberto. 2021a. Retrieval of dust properties from spectral snow reflectance measurements. *Frontiers in Environmental Science.* https://doi.org/10.3389/fenvs.2021.644551.

Kokhanovsky, A.A., F. Iodice, L. Lelli, et al. 2021b. Retrieval of total ozone column using high spatial resolution top-of-atmosphere measurements by OLCI/S-3 in the ozone Chappuis absorption bands over bright underlying surfaces. *Journal of Quantitative Spectroscopy and Radiative Transfer,* submitted.

Liou, K.N. 1992. *Radiation and cloud processes in atmosphere.* Oxford: University Press.

Painter, T. H., et al. 2012. Dust radiative forcing in snow of the Upper Colorado River Basin: 1. A 6 year record of energy balance, radiation, and dust concentrations. Water Resources Research 48 (7): W07521. https://doi.org/10.1029/2012WR011985.

Stibal, M., et al. 2015. Different bulk and active bacterial communities in cryoconite from the margin and interior of the Greenland ice sheet. *Environmental Microbiology Reports* 7: 293–300. https://doi.org/10.1111/1758-2229.12246.

Riggs, G.A., D.K. Hall, and M.O. Román. 2016. MODIS snow products collection 6 user guide, GSFC/NASA. https://modis-snow-ice.gsfc.nasa.gov/uploads/C6_MODIS_Snow_User_Guide.pdf.

Riggs, G.A., D.K. Hall, and M.O. Román. 2017. Overview of NASA's MODIS and Visible Infrared Imaging Radiometer Suite (VIIRS) snow-cover earth system data records. Earth System Sci*ence* Data 9 (2): 765–777. https://www.earth-syst-sci-data-discuss.net/essd-2017-25/.

Zege, E.P., M.P. Znachenok, and I.L. Katsev. 1980. Determination of the optical characteristics of scattering layers from diffuse reflection and transmission. *Journal of Applied Spectroscopy* 33: 1143–1148. https://doi.org/10.1007/BF00608394.

Appendix

A.1 Complex Refractive Index of Ice

In this Appendix we present the spectral complex refractive index of ice $m = n-i\chi$. It must be pointed out that the imaginary part of ice refractive index is very small in the range 300–600 nm (in the optical range). Therefore, for many purposes, one may assume that it is equal to zero. However, the precise value of χ is needed for several applications including the determination of snow and ice impurity load using the measurements in the visible region of the electromagnetic spectrum, visible light

Table A.1 The spectral dependence of complex ice refractive index $m = n-i\chi$ (Warren and Brandt 2008)

λ, μm	n	χ
3.000E-001	1.3339	2.0E-011
3.500E-001	1.3249	2.0E-011
3.900E-001	1.3203	2.0E-011
4.000E-001	1.3194	2.365E-011
4.100E-001	1.3185	2.669E-011
4.200E-001	1.3177	3.135E-011
4.300E-001	1.3170	4.140E-011
4.400E-001	1.3163	6.268E-011
4.500E-001	1.3157	9.239E-011
4.600E-001	1.3151	1.325E-010
4.700E-001	1.3145	1.956E-010
4.800E-001	1.3140	2.861E-010
4.900E-001	1.3135	4.172E-010
5.000E-001	1.3130	5.889E-010
5.100E-001	1.3126	8.036E-010
5.200E-001	1.3121	1.076E-009

(continued)

© Springer Nature Switzerland AG 2021
A. Kokhanovsky, *Snow Optics*,
https://doi.org/10.1007/978-3-030-86589-4

Table A.1 (continued)

λ, μm	n	χ
5.300E-001	1.3117	1.409E-009
5.400E-001	1.3114	1.813E-009
5.500E-001	1.3110	2.289E-009
5.600E-001	1.3106	2.839E-009
5.700E-001	1.3103	3.461E-009
5.800E-001	1.3100	4.159E-009
5.900E-001	1.3097	4.930E-009
6.000E-001	1.3094	5.730E-009
6.100E-001	1.3091	6.890E-009
6.200E-001	1.3088	8.580E-009
6.300E-001	1.3085	1.040E-008
6.400E-001	1.3083	1.220E-008
6.500E-001	1.3080	1.430E-008
6.600E-001	1.3078	1.660E-008
6.700E-001	1.3076	1.890E-008
6.800E-001	1.3073	2.090E-008
6.900E-001	1.3071	2.400E-008
7.000E-001	1.3069	2.900E-008
7.100E-001	1.3067	3.440E-008
7.200E-001	1.3065	4.030E-008
7.300E-001	1.3062	4.300E-008
7.400E-001	1.3060	4.920E-008
7.500E-001	1.3059	5.870E-008
7.600E-001	1.3057	7.080E-008
7.700E-001	1.3055	8.580E-008
7.800E-001	1.3053	1.020E-007
7.900E-001	1.3051	1.180E-007
8.000E-001	1.3049	1.340E-007
8.100E-001	1.3047	1.400E-007
8.200E-001	1.3046	1.430E-007
8.300E-001	1.3044	1.450E-007
8.400E-001	1.3042	1.510E-007
8.500E-001	1.3040	1.830E-007
8.600E-001	1.3039	2.150E-007
8.700E-001	1.3037	2.650E-007
8.800E-001	1.3035	3.350E-007
8.900E-001	1.3033	3.920E-007
9.000E-001	1.3032	4.200E-007

(continued)

Table A.1 (continued)

λ, μm	n	χ
9.100E-001	1.3030	4.440E-007
9.200E-001	1.3028	4.740E-007
9.300E-001	1.3027	5.110E-007
9.400E-001	1.3025	5.530E-007
9.500E-001	1.3023	6.020E-007
9.600E-001	1.3022	7.550E-007
9.700E-001	1.3020	9.260E-007
9.800E-001	1.3019	1.120E-006
9.900E-001	1.3017	1.330E-006
1.000E + 000	1.3015	1.620E-006
1.010E + 000	1.3014	2.000E-006
1.020E + 000	1.3012	2.250E-006
1.030E + 000	1.3010	2.330E-006
1.040E + 000	1.3009	2.330E-006
1.050E + 000	1.3007	2.170E-006
1.060E + 000	1.3005	1.960E-006
1.070E + 000	1.3003	1.810E-006
1.080E + 000	1.3002	1.740E-006
1.090E + 000	1.3000	1.730E-006
1.100E + 000	1.2998	1.700E-006
1.110E + 000	1.2997	1.760E-006
1.120E + 000	1.2995	1.820E-006
1.130E + 000	1.2993	2.040E-006
1.140E + 000	1.2991	2.250E-006
1.150E + 000	1.2990	2.290E-006
1.160E + 000	1.2988	3.040E-006
1.170E + 000	1.2986	3.840E-006
1.180E + 000	1.2984	4.770E-006
1.190E + 000	1.2982	5.760E-006
1.200E + 000	1.2980	6.710E-006
1.210E + 000	1.2979	8.660E-006
1.220E + 000	1.2977	1.020E-005
1.230E + 000	1.2975	1.130E-005
1.240E + 000	1.2973	1.220E-005
1.250E + 000	1.2971	1.290E-005
1.260E + 000	1.2969	1.320E-005
1.270E + 000	1.2967	1.350E-005
1.280E + 000	1.2965	1.330E-005

(continued)

Table A.1 (continued)

λ, μm	n	χ
1.290E + 000	1.2963	1.320E-005
1.300E + 000	1.2961	1.320E-005
1.310E + 000	1.2959	1.310E-005
1.320E + 000	1.2957	1.320E-005
1.330E + 000	1.2955	1.320E-005
1.340E + 000	1.2953	1.340E-005
1.350E + 000	1.2951	1.390E-005
1.360E + 000	1.2949	1.420E-005
1.370E + 000	1.2946	1.480E-005
1.380E + 000	1.2944	1.580E-005
1.390E + 000	1.2941	1.740E-00
1.400E + 000	1.2939	1.980E-005
1.410E + 000	1.2937	3.442E-005
1.420E + 000	1.2934	5.959E-005
1.430E + 000	1.2931	1.028E-004
1.440E + 000	1.2929	1.516E-004
1.449E + 000	1.2927	2.030E-004
1.460E + 000	1.2924	2.942E-004
1.471E + 000	1.2921	3.987E-004
1.481E + 000	1.2920	4.941E-004
1.493E + 000	1.2918	5.532E-004
1.504E + 000	1.2916	5.373E-004
1.515E + 000	1.2914	5.143E-004
1.527E + 000	1.2912	4.908E-004
1.538E + 000	1.2909	4.594E-004
1.563E + 000	1.2903	3.858E-004
1.587E + 000	1.2897	3.105E-004
1.613E + 000	1.2890	2.659E-004
1.650E + 000	1.2879	2.361E-004
1.680E + 000	1.2870	2.046E-004
1.700E + 000	1.2863	1.875E-004
1.730E + 000	1.2853	1.650E-004
1.760E + 000	1.2843	1.522E-004
1.800E + 000	1.2828	1.411E-004
1.830E + 000	1.2816	1.302E-004
1.840E + 000	1.2811	1.310E-004
1.850E + 000	1.2807	1.339E-004

(continued)

Table A.1 (continued)

λ, μm	n	χ
1.855E + 000	1.2805	1.377E-004
1.860E + 000	1.2802	1.432E-004
1.870E + 000	1.2797	1.632E-004
1.890E + 000	1.2788	2.566E-004
1.905E + 000	1.2780	4.081E-004
1.923E + 000	1.2771	7.060E-004
1.942E + 000	1.2762	1.108E-003
1.961E + 000	1.2756	1.442E-003
1.980E + 000	1.2750	1.614E-003
2.000E + 000	1.2744	1.640E-003
2.020E + 000	1.2736	1.566E-003
2.041E + 000	1.2728	1.458E-003
2.062E + 000	1.2718	1.267E-003
2.083E + 000	1.2707	1.023E-003
2.105E + 000	1.2694	7.586E-004
2.130E + 000	1.2677	5.255E-004
2.150E + 000	1.2663	4.025E-004
2.170E + 000	1.2648	3.235E-004
2.190E + 000	1.2633	2.707E-004
2.220E + 000	1.2609	2.228E-004
2.240E + 000	1.2591	2.037E-004
2.245E + 000	1.2587	2.026E-004
2.250E + 000	1.2582	2.035E-004
2.260E + 000	1.2573	2.078E-004
2.270E + 000	1.2564	2.171E-004
2.290E + 000	1.2545	2.538E-004
2.310E + 000	1.2525	3.138E-004
2.330E + 000	1.2504	3.858E-004
2.350E + 000	1.2482	4.591E-004
2.370E + 000	1.2459	5.187E-004
2.390E + 000	1.2435	5.605E-004
2.410E + 000	1.2409	5.956E-004
2.430E + 000	1.2382	6.259E-004
2.460E + 000	1.2337	6.820E-004
2.500E + 000	1.2270	7.530E-004
2.520E + 000	1.2232	7.685E-004
2.550E + 000	1.2169	7.647E-004
2.565E + 000	1.2135	7.473E-004

(continued)

Table A.1 (continued)

λ, μm	n	χ
2.580E + 000	1.2097	7.392E-004
2.590E + 000	1.2071	7.437E-004
2.600E + 000	1.2043	7.543E-004
2.620E + 000	1.1983	8.059E-004
2.675E + 000	1.1776	1.367E-003
2.725E + 000	1.1507	3.508E-003
2.778E + 000	1.1083	1.346E-002
2.817E + 000	1.0657	3.245E-002
2.833E + 000	1.0453	4.572E-002
2.849E + 000	1.0236	6.287E-002
2.865E + 000	1.0001	8.548E-002
2.882E + 000	0.9747	1.198E-001
2.899E + 000	0.9563	1.690E-001
2.915E + 000	0.9538	2.210E-001
2.933E + 000	0.9678	2.760E-001
2.950E + 000	0.9873	3.120E-001
2.967E + 000	1.0026	3.470E-001
2.985E + 000	1.0180	3.880E-001
3.003E + 000	1.0390	4.380E-001
3.021E + 000	1.0722	4.930E-001
3.040E + 000	1.1259	5.540E-001
3.058E + 000	1.2089	6.120E-001
3.077E + 000	1.3215	6.250E-001
3.096E + 000	1.4225	5.930E-001
3.115E + 000	1.4933	5.390E-001
3.135E + 000	1.5478	4.910E-001
3.155E + 000	1.5970	4.380E-001
3.175E + 000	1.6336	3.720E-001
3.195E + 000	1.6477	3.000E-001
3.215E + 000	1.6405	2.380E-001
3.236E + 000	1.6248	1.930E-001
3.257E + 000	1.6108	1.580E-001
3.279E + 000	1.5905	1.210E-001
3.300E + 000	1.5714	1.030E-001
3.322E + 000	1.5559	8.360E-002
3.345E + 000	1.5396	6.680E-002
3.367E + 000	1.5241	5.312E-002
3.390E + 000	1.5086	4.286E-002
3.413E + 000	1.4949	3.523E-002

(continued)

Table A.1 (continued)

λ, μm	n	χ
3.436E + 000	1.4827	2.887E-002
3.460E + 000	1.4710	2.347E-002
3.484E + 000	1.4604	1.921E-002
3.509E + 000	1.4502	1.586E-002
3.534E + 000	1.4411	1.326E-002
3.559E + 000	1.4328	1.130E-002
3.624E + 000	1.4146	8.146E-003
3.732E + 000	1.3924	6.672E-003
3.775E + 000	1.3850	6.966E-003
3.847E + 000	1.3750	8.248E-003
3.969E + 000	1.3623	1.112E-002
4.099E + 000	1.3526	1.471E-002
4.239E + 000	1.3447	1.867E-002
4.348E + 000	1.3406	2.411E-002
4.387E + 000	1.3401	2.656E-002
4.444E + 000	1.3412	2.990E-002
4.505E + 000	1.3444	3.179E-002
4.547E + 000	1.3473	3.090E-002
4.560E + 000	1.3482	3.007E-002
4.580E + 000	1.3491	2.883E-002
4.719E + 000	1.3470	1.940E-002
4.904E + 000	1.3379	1.347E-002
5.000E + 000	1.3325	1.240E-002

penetration in snow and ice, estimation of actinic flux, etc. The results for the ice refractive index by different authors are given in Tables A.1 and A.2. The uncertainty of measurements can be understood from Table A.3 and Fig. A.1.

Table A.2 The spectral dependence of imaginary part of ice refractive index (Picard et al. 2016)

λ, μm	χ
0.320	8.3888e-10
0.325	7.9573e-10
0.330	7.9094e-10
0.335	7.7136e-10
0.340	7.4734e-10
0.345	7.1432e-10
0.350	6.9910e-10
0.355	6.7807e-10
0.360	6.6378e-10
0.365	6.5530e-10
0.370	6.4780e-10
0.375	6.3994e-10
0.380	6.4204e-10
0.385	6.2780e-10
0.390	6.3396e-10
0.395	6.1615e-10
0.400	6.2751e-10
0.405	6.0610e-10
0.410	5.8487e-10
0.415	5.7155e-10
0.420	5.7645e-10
0.425	5.8649e-10
0.430	5.9724e-10
0.435	6.1960e-10
0.440	6.3743e-10
0.445	6.6116e-10
0.450	6.8956e-10
0.455	7.1863e-10
0.460	7.5197e-10
0.465	7.8996e-10
0.470	8.4021e-10
0.475	8.9213e-10
0.480	9.4687e-10
0.485	1.0089e-09
0.490	1.0791e-09
0.495	1.1593e-09
0.500	1.2457e-09

(continued)

Table A.2 (continued)

λ, μm	χ
0.505	1.3446e-09
0.510	1.4568e-09
0.515	1.5652e-09
0.520	1.6960e-09
0.525	1.8334e-09
0.530	1.9857e-09
0.535	2.1573e-09
0.540	2.3470e-09
0.545	2.5657e-09
0.550	2.8008e-09
0.555	3.0635e-09
0.560	3.3461e-09
0.565	3.6448e-09
0.570	3.9598e-09
0.575	4.3020e-09
0.580	4.6519e-09
0.585	5.0425e-09
0.590	5.4761e-09
0.595	5.9682e-09
0.600	5.7300e-09

Table A.3 The imaginary part of ice refractive index (multiplied by 10^9) at selected wavelengths in the visible from different sources. The differences in the spectral range 0.35–0.55 μm are particularly large χ (Picard et al. 2016)

λ, μm	χ (Warren 1984)	χ (Warren and Brandt 2008)	χ (Picard et al. 2016)
0.35	3.675	0.020	0.700
0.40	2.710	0.024	0.628
0.45	1.540	0.092	0.690
0.50	1.910	0.589	1.246
0.55	3.110	2.289	2.801
0.60	5.730	5.730	5.730

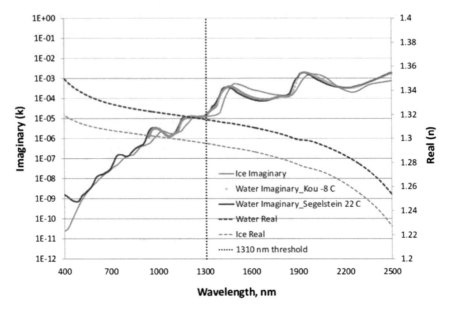

Fig. A.1 Real (n) and imaginary (k) parts of ice (blue lines) and water (red lines) refractive indices (Gallet et al. 2014). There are several wavelengths, where the values of the imaginary parts of ice and water refractive indices coincide. One of them is located close at 1300 nm. The real part of ice refractive index is systematically smaller as compared to the value of n for water

A.2 The Polarization Characteristics of Singly Scattered Light

The Stokes vector of the reflected light beam $\vec{S}$ in the single scattering approximation can be presented in the following form (Hovenier et al. 2004):

$$\vec{S} = \hat{M}(\pi - \varphi_2)\hat{P}(\theta)\hat{M}(-\varphi_1)\vec{S}_0, \qquad (A.2.1)$$

where $\hat{P}(\theta)$ is the 4×4 phase matrix, $\hat{M}(\phi)$ is the 4×4 rotational matrix, $\vec{S}_0$ is the Stokes vector of the incident solar light. The rotational matrices in Eq. (A.2.1) are needed because the Stokes vectors $\vec{S}_0(\vec{S})$ are defined with respect to the meridional plane containing the direction of light incidence (observation) and normal to the scattering layer. The phase matrix is defined with respect to the scattering plane containing directions of incident and scattered light. It follows for the rotational matrix:

$$\hat{M}(\phi) = \begin{pmatrix} 1 & 0 & 0 & 0 \\ 0 & \cos 2\phi & -\sin 2\phi & 0 \\ 0 & \sin 2\phi & \cos 2\phi & 0 \\ 0 & 0 & 0 & 1 \end{pmatrix}. \qquad (A.2.2)$$

The angles φ_1 and φ_2 can be derived from the following equations (Hovenier et al. 2004):

$$\cos\varphi_1 = \frac{-\mu_0 + \mu\gamma}{s\sqrt{(1-\mu^2)(1-\gamma^2)}}, \quad \cos\varphi_2 = -\frac{-\mu + \mu_0\gamma}{s\sqrt{(1-\mu_0^2)(1-\gamma^2)}}, \qquad (A.2.3)$$

where μ_0 is the cosine of the incidence zenith angle, μ is the cosine of the viewing zenith angle, $s = \text{sgn}(\psi - \pi)$, ψ is the relative azimuthal angle and

$$\gamma = \mu\mu_0 + \sqrt{(1-\mu_0^2)(1-\mu^2)}\sin\psi \qquad (A.2.4)$$

is the cosine of scattering angle. The phase matrix has been introduced in Chapter 3. It follows at $\mu_0 = 1$:

$$\gamma = \mu; \ \cos\varphi_1 = \cos\varphi_2 = 1. \qquad (A.2.5)$$

Then rotational matrix coincides with the unity matrix. Assuming that the incident light is unpolarized, one derives from Eq. (A.2.1):

$$\begin{pmatrix} I \\ Q \\ U \\ V \end{pmatrix} = \begin{pmatrix} P_{11}I_0 \\ \cos(2\varphi_2)P_{12}I_0 \\ -\sin(2\varphi_2)P_{12}I_0 \\ 0 \end{pmatrix}. \qquad (A.2.6)$$

Therefore, it follows for the degree of linear polarization:

$$p = \sqrt{p_q^2 + p_u^2}, \qquad (A.2.7)$$

where

$$p_q = -\frac{Q}{U} \equiv -\cos 2\varphi_2 \frac{P_{12}}{P_{11}} \qquad (A.2.8)$$

and

$$p_u = \frac{U}{I} \equiv -\sin 2\varphi_2 \frac{P_{12}}{P_{11}}. \qquad (A.2.9)$$

One can see that $p_u = tg(2\varphi_2)p_q, p = p_q$ for the case considered. It follows that in the principal plane: $\varphi_2 = 0$ or π, $tg(2\varphi_2) = 0$ and, therefore, $p_q = -p_{12} \equiv -\frac{P_{12}}{P_{11}}$ and $p_u = 0$.

A.3   The Simplified Radiative Transfer Model

We shall present the top-of-atmosphere (TOA) reflectance for atmosphere-underlying snow system in the following way (Liou 1992, 2002; Kokhanovsky 2020):

$$R_{TOA} = RT_g, \quad R = R_a + \frac{T_a f r_s}{1 - r_a r_s}, \tag{A.3.1}$$

where R is atmospheric reflectance of an idealized gas—free terrestrial atmosphere, R_a the atmospheric (with account for molecular scattering and aerosol absorption and scattering processes) contribution to the TOA reflectance, r_a is the spherical albedo of the atmosphere, r_s is the bottom-of-atmosphere snow spherical albedo, T_a is the total atmospheric transmittance from the top-of-atmosphere to the underlying surface and back to the satellite position (without account for the gaseous components of the atmosphere), T_g is the atmospheric transmittance due to the gaseous components of the atmosphere.

In the case of Lambertian underlying surfaces, the underlying surface reflectance does not depend on solar and viewing observation directions and Eq. (A.3.1) is valid with $f = 1$ and $r_s = R_s$, where R_s is underlying Lambertian surface reflectance. The snow is not exactly Lambertian reflector, therefore, we introduce the factor f in Eq. (A.3.1) to partially account for the non-Lambertian character of snow reflectance. We shall assume that

$$f = \frac{R_{snow}}{r_s}, \tag{A.3.2}$$

where R_{snow} is the snow reflection function.

The top of atmosphere reflectance R_a for a clean atmosphere can be presented in the following way using the Sobolev approximation derived from the scalar radiative transfer equation (RT) for non-absorbing media (Sobolev 1972; Kokhanovsky et al. 2020):

$$R_a = R_{ss} + R_{ms}, \tag{A.3.3}$$

where single scattering contribution

$$R_{ss} = M(\tau) p(\theta) \tag{A.3.4}$$

and multiple light scattering contribution is approximated as

$$R_{ms} = 1 + M(\tau) q(\mu_0, \mu) - \frac{N(\tau)}{4 + 3(1 - g)\tau}, \tag{A.3.5}$$

where

$$M(\tau) = \frac{1 - e^{-m\tau}}{4(\mu_0 + \mu)}, \ N(\tau) = f(\mu_0)f(\mu), \tag{A.3.6}$$

$$f(\mu) = 1 + \frac{3}{2}\mu + \left(1 - \frac{3}{2}\mu\right)e^{-\frac{\tau}{\mu}}, \ m = \mu_0^{-1} + \mu^{-1}, \tag{A.3.7}$$

$$q(\mu_0, \mu) = 3(1 + g)\mu_0\mu - 2(\mu_0 + \mu). \tag{A.3.8}$$

Here, μ_0 is the cosine of the solar zenith angle (SZA), μ is the cosine of the viewing zenith angle (VZA), θ is the scattering angle defined as $\cos\theta = -\mu_0\mu + s_0 s \cos\varphi$, where φ is the relative azimuthal angle (equal to 180 degrees minus OLCI relative azimuthal angle), s_0 is the sine of the SZA, s is the sine of the VZA, τ is the atmospheric optical thickness, $p(\theta)$ is the phase function, g is the asymmetry parameter determined by the following expression:

$$g = \frac{1}{2}\int_0^\pi p(\theta)\sin\theta\cos\theta d\theta. \tag{A.3.9}$$

The approximate account for aerosol absorption effects can be performed multiplying R_{ss} by the single scattering albedo ω_0. The accuracy of Eqs. (A.3.1–A.3.3) can be further improved using the truncation approximation as discussed by Katsev et al. (2010).

The phase function is presented in the following form (assuming that absorption effects can be ignored):

$$p(\theta) = (\tau_{aer}p_{aer}(\theta) + \tau_m p_m(\theta))/\tau, \tag{A.3.10}$$

where

$$p_{aer}(\theta) = cp_{aer,1}(\theta) + (1 - c)p_{aer,2}(\theta), \tag{A.3.11}$$

$$p_{aer,1}(\theta) = \frac{1 - g_{aer,1}^2}{\left(1 - 2g_{aer,1}\cos\theta + g_{aer,1}^2\right)^{\frac{3}{2}}}, \ p_{aer,2}(\theta) = \frac{1 - g_{aer,2}^2}{\left(1 - 2g_{aer,2}\cos\theta + g_{aer,2}^2\right)^{\frac{3}{2}}}, \tag{A.3.12}$$

$$g_{aer,1} = 0.8, \ g_{aer2} = -0.45, \tag{A.3.13}$$

$$c = \frac{g_a - g_{aer,2}}{g_{aer,1} - g_{aer,2}}, \tag{A.3.14}$$

$$p_m(\theta) = \frac{3}{4}(1 + \cos^2\theta) \tag{A.3.15}$$

is the molecular scattering phase function. Therefore, it follows for the asymmetry parameter:

$$g = \frac{\tau_{aer}}{\tau_{mol} + \tau_{aer}} g_a. \tag{A.3.16}$$

The parameter g_a varies with the location, time, aerosol, type, etc. We shall assume that it can be approximated by the following equation:

$$g_a = g_0 + g_1 e^{-\frac{\lambda}{\lambda_0}}. \tag{A.3.17}$$

The coefficients in this equation (as derived from multiple year AERONET observations over Greenland (Kokhanovsky et al. 2020) are:

$$g_0 = 0.5263, \; g_1 = 0.4627, \; \lambda_0 = 0.4685 \mu m. \tag{A.3.18}$$

The transmission function $T(\mu_0, \mu)$ is approximated as follows:

$$T(\mu_0, \mu) = t^m, \tag{A.3.19}$$

where t is calculated using the following approximation (Katsev et al. 2010):

$$t = e^{-B\tau}, \tag{A.3.20}$$

where

$$B = \frac{1}{2} \int_{\frac{\pi}{2}}^{\pi} p(\theta) \sin \theta d\theta \tag{A.3.21}$$

is the so – called backscattering fraction and τ is the atmospheric optical thickness. It follows from Eqs. (A3.21 and A3.10):

$$B = (\tau_{aer} B_a + \tau_m B_m)/\tau, \tag{A.3.22}$$

where $B_m = 0.5$,

$$B_a = c B_{aer}(g_{aer,1}) + (1 - c) B_{aer}(g_{aer,2}), \tag{A.3.23}$$

$$B_{aer}(g) = \frac{1 - g}{2g} \left(\frac{1 + g}{\sqrt{1 + g^2}} - 1 \right). \tag{A.3.24}$$

Table A.3.1 The polynomial expansion coefficients of the parameters given in Eq. (A.3.26). The parameter M is given by the following equation: $M = \sum\limits_{s=0}^{3} M_s g^s$ (and similar for other parameters)

s	0	1	2	3
M_s	0.18016	−0.18229	0.15535	−0.14223
N_s	0.58331	−0.50662	−0.09012	0.020700
D_s	0.21475	−0.10000	0.13639	−0.21948
ξ_s	0.16775	−0.06969	0.08093	−0.08903
κ_s	1.09188	0.08994	0.49647	−0.75218

It follows from Eq. (A.3.24) that $B_{aer}(1) = 0$ as it should be. Also we have from Eq. (A.3.24) at small values of g:

$$B_{aer}(g) = \frac{1}{2} + \frac{g(g^2 - 3)}{2(1 + g^2 + (1 - g^2)\sqrt{1 + g^2})}. \qquad (A.3.25)$$

Therefore, it follows: $B_{aer}(0) = 0.5$ as it should be for the symmetric phase functions (e.g., for molecular scattering).

The atmospheric spherical albedo r_a is found using the approximation proposed by Kokhanovsky et al. (2005):

$$r_a = \left(M e^{-\frac{\tau}{\xi}} + N e^{-\frac{\tau}{\kappa}} + D \right) \tau. \qquad (A.3.26)$$

The coefficients of polynomial expansions of all coefficients (M, N, D, ξ, κ) in Eq. (A.3.26) with respect to the value of the asymmetry parameter g are given in next Table A.3.1 (Kokhanovsky et al. 2005).

One can see that the functions discussed above depend on the atmospheric optical thickness, which can be presented in the following form:

$$\tau(\lambda) = \tau_{mol}(\lambda) + \tau_{aer}(\lambda). \qquad (A.3.27)$$

The molecular optical thickness can be approximated as (Iqbal 1983):

$$\tau_{mol}(\lambda) = q\lambda^{-\upsilon} \qquad (A.3.28)$$

at the normal pressure p_0 and temperature T_0. Here, $q = 0.008735$, $\upsilon = 4.08$, and the wavelength is in microns. The value of molecular optical thickness at another pressure level p can be derived using the following expression: $\tau_{mol}(\lambda) = \hat{p}\tau_m(\lambda)$, where $\hat{p} = \frac{p}{p_0}$, p is the site pressure, $p_0 = 1013.25mb$. The site pressure is calculated as: $p = p_0 \exp\left(-\frac{z}{H}\right)$. Here z is the height of the underlying surface and $H = 7.64\,km$

is the scale height. The molecular optical thickness at Arctic and Antarctic sites is discussed in detail by Tomasi and Petkov (2015).

The aerosol optical thickness can be presented as

$$\tau_{aer}(\lambda) = \beta\left(\frac{\lambda}{\lambda_0}\right)^{-\alpha}, \tag{A.3.29}$$

where $\lambda_0 = 0.4\mu m$, $(\alpha, \beta = \tau_{aer}(\lambda_0))$ are the Angström parameters. One can assume that $\alpha = 1.6$ (Six et al. 2005).

As far as gaseous transmission is of concern, one can use the following exponential approximation:

$$T_g(\lambda) = \exp\left(-A\tau_g(\lambda)\right), \tag{A.3.30}$$

where A is the air mass factor (Rozanov and Rozanov 2010) and $\tau_g(\lambda)$ is the gaseous absorption optical thickness.

The air mass factor can be approximated by the following equation for the case of ozone (Iqbal 1983):

$$A = \frac{1+s}{\sqrt{2s + \mu_0^2}} + \frac{1+s}{\sqrt{2s + \mu^2}}, \tag{A.3.31}$$

$s = H/a$, H is the height of ozone layer (assumed to be equal 22 km), $a = 6370$ km is the radius of the Earth. The account for the latitudinal change of the ozone layer height can be performed using the following empirical relationship (Savastiouk and McErloy 2004): $H = x - 0.1y$, where y is the latitude in degrees without regard for the sign (e.g., $90°$ for the South and North Poles) and $x = 26$ km.

References

Gallet, J.-C., F. Domine, and M. Dumont. 2014. Measuring the specific surface area of wet snow using 1310 nm reflectance, *The Cryosphere* 8: 1139–1148. https://doi.org/10.5194/tc-8-1139-2014.

Hovenier, J.W., C. van der Mee, and H. Domke. 1971. *Transfer of polarized light in planetary atmospheres*. Dordrecht: Kluwer.

Iqbal, M. 1983. *An Introduction to Solar Radiation*, New York: Academic Press, p. 101.

Katsev, I.L., A.S. Prikhach, E.P. Zege, J.O. Grudo, and A.A. Kokhanovsky. 2010: Speeding up the aerosol optical thickness retrieval using analytical solutions of radiative transfer theory. *Atmospheric Measurement Techniques* 3: 1403–1422. https://doi.org/10.5194/amt-3-1403-2010.

Kokhanovsky A.A., B. Mayer, and V.V. Rozanov. 2005. A parameterization of the diffuse transmittance and reflectance for aerosol remote sensing problems. *Atmospheric Research* 73 (1): 37–43.

Kokhanovsky, A., J.E. Box, B. Vandecrux, K.D. Mankoff, M. Lamare, A. Smirnov, and M. Kern. 2020. The determination of snow albedo from satellite measurements using fast atmospheric correction technique. *Remote Sensing* 12: 234.

Liou, K.N. 1992. *Radiation and cloud processes in atmosphere.* Oxford: Oxford University Press.

Liou, K.N. 2002. *An Introduction to Atmospheric Radiation.* N. Y. : Academic Press.

Picard, G., Q. Libois, and L. Arnaud. 2016. Refinement of the ice absorption spectrum in the visible using radiance profile measurements in Antarctic snow. *The Cryosphere* 10: 2655–2672. https://doi.org/10.5194/tc-10-2655-2016.

Rozanov, V.V., and A.V. Rozanov. 2010: Differential optical absorption spectroscopy (DOAS) and air mass factor concept for a multiply scattering vertically inhomogeneous medium: theoretical consideration. *Atmospheric Measurement Techniques* 3: 751–780. https://doi.org/10.5194/amt-3-751-2010.

Savastiouk, V., and C. T. McErloy. 2004. Calculating air mass factors for ozone and Rayleigh air mass factor calculations for ground-based spectrometers. *Proceedings. of Quadrennial Ozone Symposium*, Kos, Greece. https://doi.org/10.13140/2.1.3553.7284.

Six D., M. Fily, L. Blarel, and P. Goloub. 2005. First aerosol optical thickness measurements at Dome C (east Antarctica), summer season 2003–2004. *Atmospheric Enviroment* 39: 5041-5050.

Tomasi, C., and B.H. Petkov. 2015. Specral calculations of Rayleigh—scattering optical depth at Arctic and Antarctic sites using a two–term algorithm. *Journal of Geophysical Research.* https://doi.org/10.1002/2015JD023575.

Sobolev, V.V. 1972. *Light scattering in planetary atmospheres.* Moscow: Nauka.

Warren, S.G. 1984. Optical constants of ice from the ultraviolet to the microwave. *Applied Optics* 23: 1206-1225.

Warren, S., and R.E. Brand. 2008. Optical constants of ice from the ultraviolet to the microwave: a revised compilation, *Journal of Geophysical Research* 113: D14. https://doi.org/10.1029/2007JD009744.

Printed in the United States
by Baker & Taylor Publisher Services